Synthesis and Structural Characterization of Arsinoamide and its Metal Complexes

I0131842

Synthesis and Structural Characterization of Arsinoamide and its Metal Complexes

Zur Erlangung des akademischen Grades eines

DOKTORS DER NATURWISSENSCHAFTEN

(Dr.rer.nat.)

der KIT-Fakultät für Chemie und Biowissenschaften

des Karlsruher Instituts für Technologie (KIT)

vorgelegte

DISSERTATION

von

M.S.-Chem. Xiao Chen

aus

Yongcheng, Henan, China

KIT-Dekan: Prof. Dr. Reinhard Fischer

Referent: Prof. Dr. Peter W. Roesky

Korreferent: Prof. Dr. Annie K. Powell

Tag der mündlichen Prüfung: 08. Mai 2019

Bibliografische Information der Deutschen Nationalbibliothek

Die Deutsche Nationalbibliothek verzeichnet diese Publikation in der
Deutschen Nationalbibliografie; detaillierte bibliographische Daten sind im Internet
über http://dnb.d-nb.de abrufbar.

1. Aufl. - Göttingen: Cuvillier, 2019

   Zugl.: Karslruhe (KIT), Univ., Diss., 2019

© CUVILLIER VERLAG, Göttingen 2019

   Nonnenstieg 8, 37075 Göttingen

   Telefon: 0551-54724-0

   Telefax: 0551-54724-21

   www.cuvillier.de

1. Auflage, 2019

Gedruckt auf umweltfreundlichem, säurefreiem Papier aus nachhaltiger Forstwirtschaft.

   ISBN 978-3-7369-7034-2

   eISBN 978-3-7369-6034-3

"Everyone wants to be successful until they see what it actually takes."

Content

1 Introduction

1.1 Phosphinoamide

Phosphinoamides and their derivatives are a series of compounds featuring a phosphorus(III)-nitrogen(III) bond. Compounds containing a P-N single bond (Cl_2P-NR_2, Cl_2P-N(H)R, and RP(NR'_2)$_2$) have been first reported by Michaelis.[1,2] Later on, this series of compounds was expanded with some further examples including (CF_3)$_2$P-N(H)Me and R_2P-NR_2 (R = Me, Et, Ph).[3-5] With few exceptions, P-N bond can be formed by three general pathways (Scheme 1.1.1): 1, by the aminolysis reaction between a halogenophosphine and a primary or secondary amine, through the elimination of HCl with an organic base (eqn. 1); 2, by salt metathesis (salt elimination) between a chlorophosphine and an alkali metal amide (eqn. 2); 3, by the elimination of Me_3SiCl from the reaction between a chlorophosphine and an aminosilane (eqn. 3).[6-11] Instead of alkyl groups, aryl groups are more suitable for the synthesis of phosphinoamides. Indeed, an aryl group on the phosphorus atom can shorten the distance of the P-N bond by decreasing the electron density on the phosphorus.[12-14] In addition, quantum chemical calculation revealed that σ-acceptor substituents (*e.g.* fluorine) on the phosphorus atom could further strengthen the P-N bond. On the contrary, electron-withdrawing substituents on the nitrogen atom weaken the bond.[15]

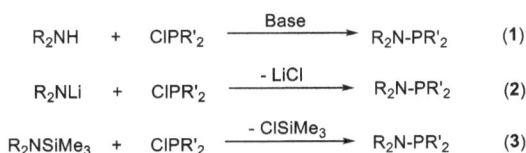

$$R_2NH \; + \; ClPR'_2 \quad \xrightarrow{\text{Base}} \quad R_2N\text{-}PR'_2 \qquad (1)$$

$$R_2NLi \; + \; ClPR'_2 \quad \xrightarrow{\text{- LiCl}} \quad R_2N\text{-}PR'_2 \qquad (2)$$

$$R_2NSiMe_3 \; + \; ClPR'_2 \quad \xrightarrow{\text{- ClSiMe}_3} \quad R_2N\text{-}PR'_2 \qquad (3)$$

Scheme 1.1.1 General pathways to form a P-N bond.

The phosphinoamide chemistry has attracted a considerable attention owing to the structural diversity of the associated complexes.[16-21] To further understand the properties of the P-N bond, Ashby synthesized the compound [(Ph_2P-NPh)Li(Et_2O)]$_2$ (Figure 1.1.1) through the deprotonation reaction of the corresponding aminophosphine with ^{n}BuLi.[22] X-ray diffraction analyses indicated the presence of Li-P bonds (2.684(3) and 3.004(4) Å). In addition, the P-N bond length (1.672(2) Å) was relatively smaller than that of a single bond (mean value of 1.70

Å), which indicates that the P-N bond distance in phosphinoamides falls between single- and double-bond distances.

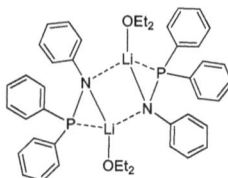

Figure 1.1.1 Molecular structure of [(Ph$_2$P-NPh)Li(OEt$_2$)]$_2$.

Quantum chemical calculations revealed that the [Ph$_2$PNPh]$^-$ anion exhibits two resonance forms: phosphinoamide (negative charge located on N) and iminophosphide (negative charge located on P) (Figure 1.1.2). A possible explanation is the stabilization of the charge α to the phosphorous by negative hyperconjugation.[23,24] Further theoretical studies indicate that in most case the negative charge is placed on the more electronegative nitrogen atom.

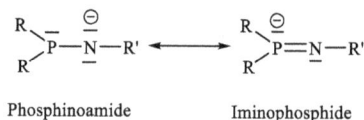

Phosphinoamide   Iminophosphide

Figure 1.1.2 Resonance forms of phosphinoamide and iminophosphide.

Two ground-state conformations, *cis* and *trans*, can be proved in the molecular structure of phosphinoamide derivatives (Figure 1.1.3). Theoretical calculations revealed that the interconversion barrier between the *cis* and *trans* isomers could be influenced by the nature of substituents. Apart from this, negative hyperconjugation could significantly affect the conformation of phosphinoamides. Specifically, phosphinoamides prefer to adopt a *cis* conformation in spite of the unfavorable steric.[25]

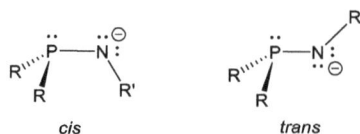

cis   *trans*

Figure 1.1.3 Ground-state conformations of phosphinoamide.

1.2 Metal complexes of phosphinoamide

Given the presence of lone pairs on both the nitrogen and phosphorus atoms, phosphinoamides and their derivatives have been widely used as ligands to establish various metal complexes.[17,26-33] In particular, the "hard" nitrogen atom could stabilize metals in high oxidation states, while the "softer" phosphorus atom is a good candidate to coordinate to metals in medium and low oxidation states. The first metal complexes of phosphinoamides, [(Ph$_2$P-NEt$_2$)CuI]$_4$ and [(Ph$_2$P-NEt$_2$)HgI$_2$], were established in 1962.[3] Metal complexes of phosphinoamide can be synthesized following two typical reaction pathways (Scheme 1.2.1): 1, through a salt metathesis reaction (**1**) between an alkali metal salt of phosphinoamide and a metal halide; 2, through an intermolecular deprotonation reaction (**2**) between a metal complex bearing a basic ligand and an aminophosphine.

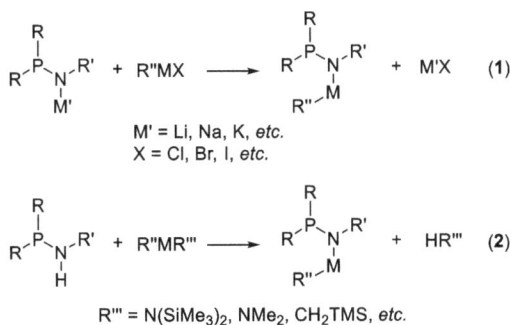

M' = Li, Na, K, *etc.*
X = Cl, Br, I, *etc.*

R''' = N(SiMe$_3$)$_2$, NMe$_2$, CH$_2$TMS, *etc.*

Scheme 1.2.1 Typical reaction pathways to synthesize phosphinoamide metal complexes.

Possessing both a hard donor atom (nitrogen) and a soft donor atom (phosphorus), phosphinoamides can exhibit various coordination modes with metal centers, acting as either monodentate, bridging or chelating ligands (Figure 1.2.1). As a monodentate ligand (**a**), the nitrogen atom directly connects to the metal center, while the lone pair on the phosphorus atom is free. As a bridging ligand (**b**), the nitrogen and phosphorus atoms connect to two different metal centers. As a chelating ligand (**c**), both the nitrogen and phosphorus atoms coordinate to one metal atom, giving a three-membered ring.

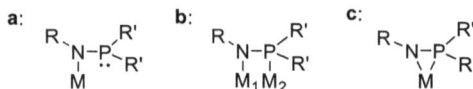

Figure 1.2.1 Coordination modes of phosphinoamide ligands.

Over the past two decades, various phosphinoamides and their metal complexes have been established and displayed promising applications. Phosphinoamide metal complexes have proved to be good catalysts in polymerization, hydroamination, hydroboration etc..[34-42] Transition metal and lanthanide complexes have especially shown interesting magnetic properties.[43-45] In addition, the biological activities could be confirmed for some phosphinoamide compounds.[46-48]

1.2.1 Lanthanide complexes of phosphinoamide

Lanthanides, abbreviated to Ln, are the series of chemical elements from lanthanum to lutetium (La, Ce, Pr, Nd, Pm, Sm, Eu, Gd, Tb, Dy, Ho, Er, Tm, Yb, and Lu). Owing to the existence of the 4f electron shell, lanthanides correspond to f-block elements. One of their most interesting features lies in the lanthanide contraction corresponding to a steady decrease in the ionic radii from lanthanum to lutetium. Owing to the presence of the 4f electron shell and the lanthanide contraction, lanthanide elements and their derivatives manifest specific spectroscopic and magnetic properties compared to transition metals and main group elements.

In 1999, three homoleptic lanthanide complexes of phosphinoamides, consisting only of three-membered metallacycles (η^2-coordinated chelating ligands), were established by our group[49], via a salt metathesis reaction, between lithium phosphinoamide and anhydrous $LnCl_3$ (Ln = Y, Yb, Lu) (Scheme 1.2.1.1). As soft Lewis bases, the four phosphorus atoms in the complexes formed very weak interactions with the lanthanide atom (hard acid), with bond distances varying from 2.885(2) to 3.040(3) Å.

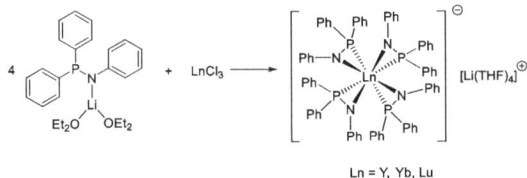

Ln = Y, Yb, Lu

Scheme 1.2.1.1 Synthesis of homoleptic phosphinoamide lanthanide complexes.

Furthermore, the palladium allyl precursor ([Pd(C$_3$H$_5$)Cl]$_2$) was utilized to establish Ln-Pd (Ln = Y, Lu) heterobimetallic complexes (Scheme 1.2.1.2).[50] The isolated heterobimetallic Ln(III)-Pd(0) complexes revealed differences compared to their corresponding Zr-Pd complexes. Indeed, the unexpected reduction of [Pd(C$_3$H$_5$)Cl]$_2$ and the formation of one neutral ligand, Ph$_2$P-N(H)Ph, were observed, leading to the formation of unanticipated bimetallic Ln(III)-Pd(0) complexes and trimetallic Pd(0)-Ln(III)-Pd(0) complexes. It was suggested that the Ph$_2$PN(H)Ph ligand comes from the decomposition of [(Ph$_2$PNPh)$_4$Ln][Li(THF)$_4$]. X-ray diffraction analyses and quantum chemical calculations illustrated the presence of the Ln-Pd interactions with bond lengths ranging from 2.9031(11) Å to 3.1860(12) Å.

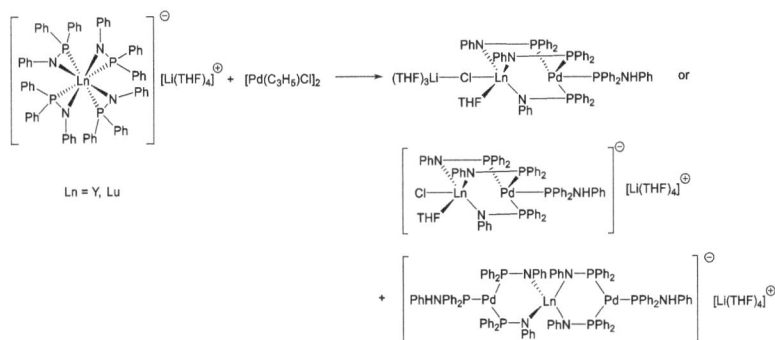

Ln = Y, Lu

Scheme 1.2.1.2 Synthesis of phosphinoamide-supported Ln-Pd complexes.

Inspired by the above Ln(III)-Pd(0) complexes, some phosphinoamide-supported Ln(III)-Pt(0) complexes have also been reported by our group, through the reaction of [(Ph$_2$PNPh)$_4$Ln][Li(THF)$_4$] (Ln = Y, Lu) with [Pt(^{t}Bu$_3$P)$_2$] (Scheme 1.2.1.3).[51] The platinum atom could displace the phosphine donors, resulting in the cleavage of the three Ln-P bonds and the formation of Ln-Pt interactions (Y-Pt: 3.0063(8) Å and Lu-Pt: 2.9523(9) Å). In addition, the presence of LiCl could influence the outcome of this reaction. Specifically, in the absence of

LiCl, one phosphinoamide anion [PPh₂NPh]⁻ chelates the lanthanide atom forming a three-membered metallacycle, while, in the presence of LiCl, one [LiCl(THF)₃] moiety is coordinated to the lanthanide ion via a bridging chloride atom.

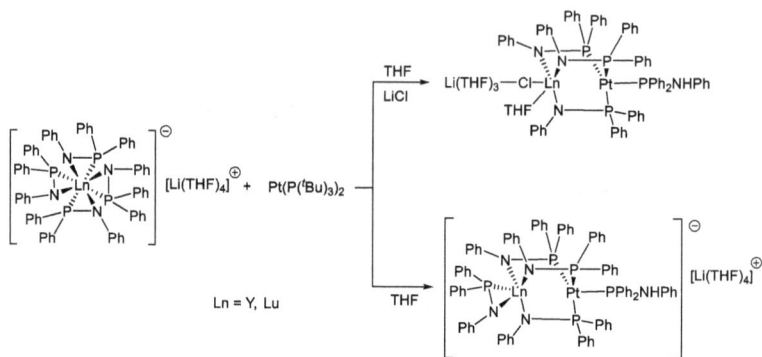

Scheme 1.2.1.3 Synthesis of phosphinoamide-supported Ln(III)-Pt(0) heterobimetallic complexes.

1.2.2 Group 4 metal complexes of phosphinoamide

Group 4 is the group of elements containing titanium (Ti), zirconium (Zr), hafnium (Hf) and rutherfordium (Rf), with a $(n-1)d^2ns^2$ electron configuration, favoring the +4 oxidation state. For titanium, the oxidation states of +2 and + 3 have been confirmed. Due to the lanthanide contraction, zirconium and hafnium have a very close atomic radius and thus manifest very similar chemical reactivity.

After the report of the first heterobimetallic phosphinoamide group 4 complexes in 1993 (Figure 1.2.2.1)[52], large amounts of phosphinoamide group 4 complexes have been established in the past decades.[53-63]

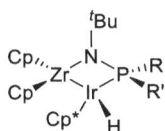

R = ethyl, H
R' = ethyl, cyclohexyl

Figure 1.2.2.1 Molecular structure of the first group 4 phosphinoamide complexes.

Reaction of TiCl$_4$ with 2 equiv. of lithium phosphinoamide resulted in the formation of their corresponding bis-substituted titanium complexes (Scheme 1.2.2.1).[64] X-ray diffraction analyses indicated the existence of Ti-P interactions with distances ranging from 2.4223(9) to 2.476(2) Å. To investigate the possibility to form Ti-Pt interactions, different platinum precursors were utilized to react with the bis(phosphinoamide)titanium complex. Both X-ray diffraction analyses and variable-temperature NMR experiments confirmed the existence and fluxionality of the Pt-Ti dative bonds (bond distances of 2.6860(11) to 2.8358(8) Å).

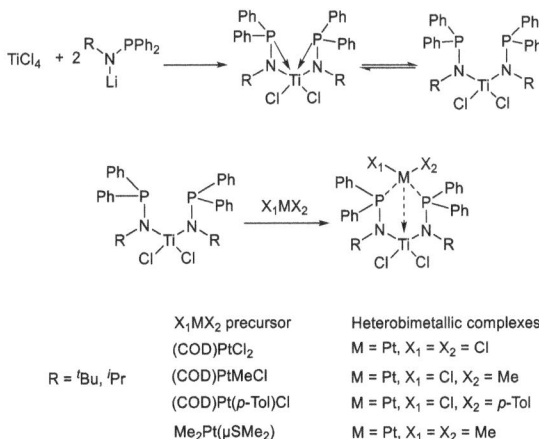

X$_1$MX$_2$ precursor	Heterobimetallic complexes
(COD)PtCl$_2$	M = Pt, X$_1$ = X$_2$ = Cl
(COD)PtMeCl	M = Pt, X$_1$ = Cl, X$_2$ = Me
(COD)Pt(p-Tol)Cl	M = Pt, X$_1$ = Cl, X$_2$ = p-Tol
Me$_2$Pt(μSMe$_2$)	M = Pt, X$_1$ = X$_2$ = Me

R = ^{t}Bu, ^{i}Pr

Scheme 1.2.2.1 Synthesis of phosphinoamide-supported Ti/Pt complexes.

In 2005, chiral phosphinoamide ligands were introduced into the group 4 coordination chemistry by our group (Scheme 1.2.2.2).[65] Salt metathesis reactions of the chiral lithium phosphinoamides with Cp$_2$ZrCl$_2$ led to the formation of the corresponding mono-substituted zirconium complexes. X-ray diffraction analyses revealed that the phosphorus atom is bonded to the zirconium atom with bond distances of 2.652(1) Å (S) and 2.650(2) Å (R). Alternatively, the enantiomerically pure mono-substituted complex was synthesized from the neutral amine and [(PhCH$_2$)$_4$Zr], *via* an intermolecular deprotonation reaction. It was suggested that the nature of the substituents on the zirconium atom could influence the Zr-N bond distance (2.1927(14) Å (R = Cp) *vs* 2.096(2) Å (R = Bn)).

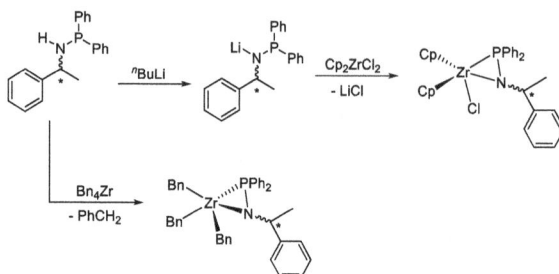

Scheme 1.2.2.2 Synthesis of chiral phosphinoamide zirconium complexes.

Nagashima reported the synthesis of tris(phosphinoamide)zirconium complexes through the reaction of ZrCl$_4$ with 3 equiv. of lithium phosphinoamides (Scheme 1.2.2.3).[66] X-ray diffraction studies indicated the existence of three P-Zr interactions with bond distances varying from 2.608(2) to 2.6656(9) Å. After further reaction with CuCl and [Mo(CO)$_3$(CH$_3$CN)$_3$] in CH$_3$CN, the corresponding phosphinoamide-supported heterobimetallic Zr-TM complexes were established with Zr-TM distances of 2.6854(6) Å (Zr-Cu) and 2.9741(5) Å (Zr-Mo).

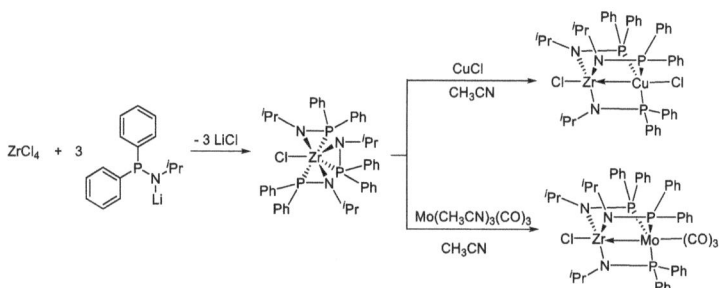

Scheme 1.2.2.3 Synthesis of the tris(phosphinoamide)zirconium complex and its heterobimetallic Zr/Cu and Zr/Mo complexes.

In 2014, Zr/Co heterobimetallic complexes bridged by phosphinoamides were crystallographically authenticated by Thomas.[67] CoI$_2$ was treated with [(^{i}Pr$_2$PNXyl)$_3$ZrCl] to generate the desired Zr/Co complex. After further reduction with 2.5 equiv. of Na/Hg, the expected phosphinoamide-supported heterobimetallic complex was obtained with a Zr-Co bond distance of 2.3778(5) Å (Scheme 1.2.2.4).

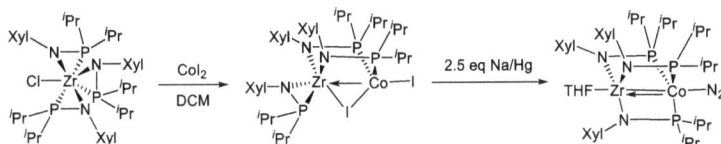

Scheme 1.2.2.4 Synthesis of phosphinoamide-supported Zr/Co complex.

To investigate the influence of the electron properties of the substituents on the metal-metal interaction, several zirconium/platinum complexes with substituents of varying electron-releasing ability was synthesized and structurally characterized (Scheme 1.2.2.5).[68] It was suggested that electron-withdrawing ligands on the zirconium atom could decrease the Pt-Zr distance. Instead, electron rich ligands could increase this separation (2.7761(5) Å (R = Cl) vs 3.2343(3) Å (R = NMe₂)).

Thomas and coworkers suggested that the metal-to-metal interaction could be regulated in phosphinoamide-supported early/late heterobimetallic complexes.[37,69,70] However, some obstacles in this field should not be ignored, such as the deficiency of suitable ligands.

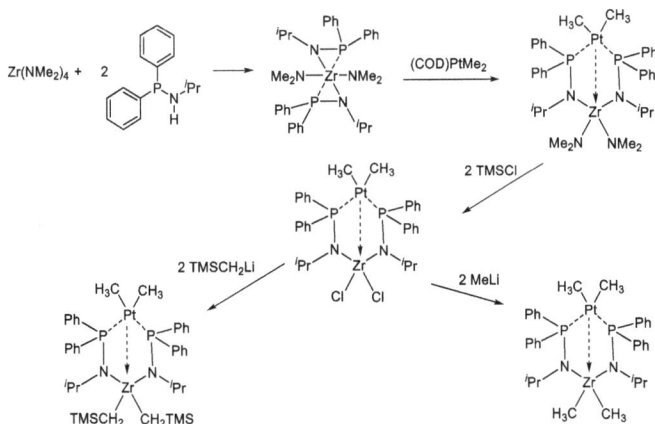

Scheme 1.2.2.5 Synthesis of phosphinoamide-supported Zr/Pt complexes.

1.2.3 Group 13 metal complexes of phosphinoamide

Group 13 contains boron (B), aluminum (Al), gallium (Ga), indium (In) and thallium (Tl). Since the possession of three electrons in the valence shell with the electron configuration of ns^2np^1, they prefer the oxidation state +3 except for the heaviest element thallium (+1 oxidation state favored).

Only one example of a group 13 metal phosphinoamide complex has been reported. In 2017, Harder synthesized the corresponding aluminum α-phosphinoamide complex from the reaction of $Ph_2PN(H)Dipp$ with $AlMe_3$ via an aminolysis reaction (Scheme 1.2.3.1).[71] This complex is dimeric bridged by two weak Al-P interactions (3.0396(12) Å). In addition, its frustrated Lewis pair reactivity towards isocyanates and CO_2 were investigated, on account of the mismatch between the aluminum (hard Lewis acid) and phosphorus atom (soft Lewis base).

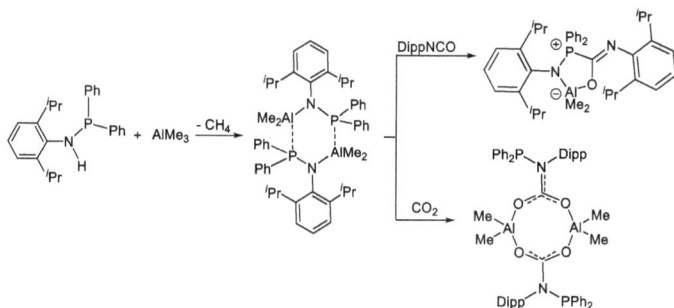

Scheme 1.2.3.1 Frustrated Lewis pair reactivity of an aluminum phosphinoamide complex.

No gallium or indium complex of phosphinoamides has ever been reported. It should be mentioned that treatment of the lithium salt of bis(phosphinimino)amines with the appropriate metal halides MX_3 (M = Al, Ga, In, X = Cl, Br), via salt metathesis, resulted in the desired dihalo complexes of bis(phosphinimino)amine with formation of a six-membered N_3P_2M ring.[72]

1.2.4 Group 14 metal complexes of phosphinoamide

Group 14 is composed of carbon (C), silicon (Si), germanium (Ge), tin (Sn) and lead (Pb). These elements possess the valence electron configuration ns^2np^2. Carbon and silicon prefer to adopt the oxidation state +4, while tin and lead prefer +2. Despite in the same group, their physical and chemical features vary significantly. Carbon is a non-metal; silicon and germanium are widely used in semiconductor industry; tin and lead are considered as metals in general.

In the oxidation state +2, the heavier carbene analogues of Si, Ge, Sn and Pb compounds are commonly called silylene, germylene, stannylene and plumbylene, respectively. Since the first report of benzamidinato-supported chlorotetrylenes ([{PhC(^{t}BuN)$_2$}ECl], E = Si, Ge, Sn), the chemistry of low-valent group 14 compounds has dramatically expanded over the last decade.[73-77] Owing to the presence of both a lone pair and a vacant p orbital, low-valent group 14 compounds possess a dual Lewis acid-base character[78-81] and thus stimulated much interest,[82-87] especially in terms of oxidative addition.[88-92] Due to their ability to split not only single bond, $e.g.$ E-H (E = H, C, N, O $etc.$), but also double bond and triple bond, $e.g.$ C=X (X = C, N, O, $etc.$) and C≡C , low-valent group 14 compounds possess a great potential application for catalytic reactions, in similar way as transition metal compounds.[93-100]

Compared with the well-established d- and f-block metal complexes of phosphinoamides, only a small number of low-valent group 14 complexes of phosphinoamides have been reported. In 2016, Khan established the synthesis of three phosphinoamide tetrylene complexes, via salt metathesis, from the reaction of [{PhC(^{t}BuN)$_2$}ECl] (E = Si, Ge, Sn) with lithium phosphinoamides (Scheme 1.2.4.1).[101,102] According to the X-ray diffraction studies, no E-P interaction or P-N bond activation was observed. Nevertheless, these compounds had a main common feature: they possessed two lone pairs (E and P), revealing a wide potential application in coordination chemistry. For example, the phosphinoamide-based silylene was utilized to investigate the aurophilic interaction.[102] After abstraction of the chloride, the desired phosphinoamide-supported complex was formed with an intermolecular Au-Au interaction of 2.865 Å (Scheme 1.2.4.2).

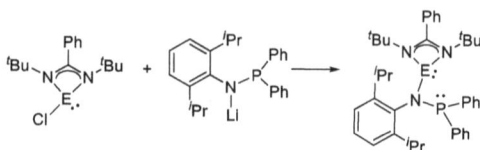

Scheme 1.2.4.1 Synthesis of phosphinoamide-supported tetrylenes.

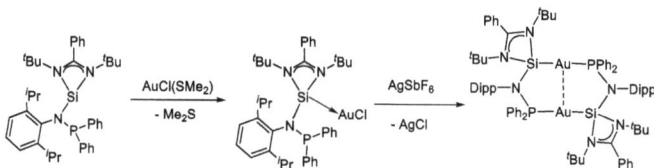

Scheme 1.2.4.2 Phosphinoamides-supported silylene-gold complexes.

According to the literature, the P-N bond activation was only achieved by the reaction of [{PhC('BuN)₂}SiCl] with the phosphinoamide DippNLiPPh₂ under harsh conditions (80 °C, 18 h), which led to the formation of the N=Si-P unit (Scheme 3.4.3).[103]

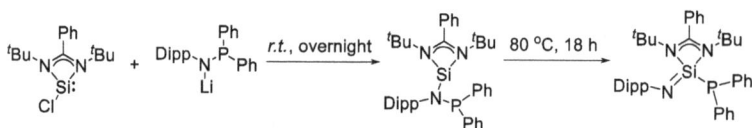

Scheme 1.2.4.3 Reaction of [{PhC('BuN)₂}SiCl] with phosphinoamide.

A handful of phosphinoamide germylene complexes have been reported. In 2015, two extremely bulky phosphinoamide germanium(II) dichloride complexes were established from the reaction of potassium phosphinoamides KL (L = Ar*N(PPh₂), Ar* = (2,4,6-'Pr)C₆H₂[C(H)Ph₂]₂) with GeCl₂·dioxane (Scheme 1.2.4.4).[104] In addition, the coordination between the phosphorus atom and GeCl₂ was observed with distances of 2.5725(8) Å and 2.5435(8) Å.

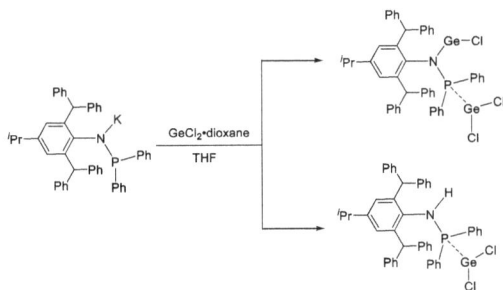

Scheme 1.2.4.4 Synthesis of extremely bulky phosphinoamides germanium(II) dichloride

complexes.

Treatment of GeCl$_2$·dioxane with 1 equiv. of lithium phosphinoamide, *via* salt metathesis, resulted in the formation of a dimeric germylene, in which a nonplanar six-membered ring (Ge$_2$N$_2$P$_2$), with a Ge-P interaction of 2.612(3) Å, could be observed (Scheme 1.2.4.5).[105] However, with 2 equiv. of lithium phosphinoamide, the desired diaminogermylene [(DippNPPh$_2$)$_2$Ge] was isolated. Nevertheless, no Ge-P interaction was detected. The consistent trend could also be observed in the pyridine substituted phosphinoamide stannylene complexes (Scheme 1.2.4.6).[106] The Sn-P interaction can only be observed in the mono-substituted stannylene complex.

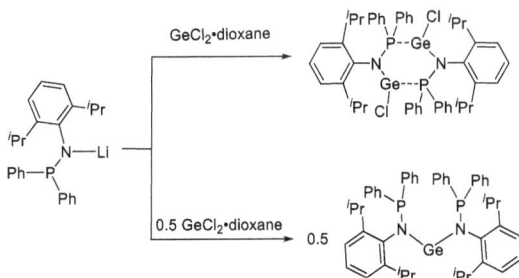

Scheme 1.2.4.5 Synthesis of phosphinoamide germylene complexes.

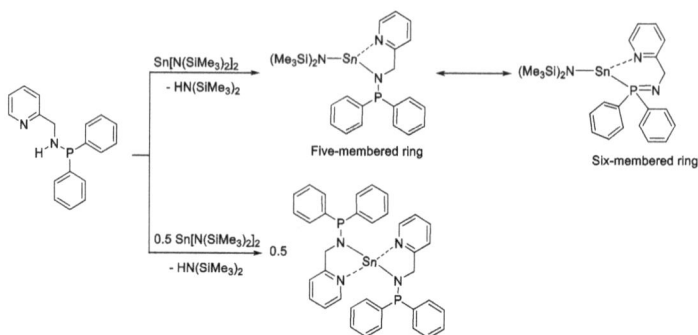

Scheme 1.2.4.6 Synthesis of phosphinoamide stannylene complexes.

1.3 Arsinoamide

Compared with the well-established coordination chemistry of phosphinoamides, arsinoamides and their metal complexes are almost unknown. In 1959, the first aminoarsanes ($(CF_3)_2AsN(H)R$, R = Me, Et) were established from the reaction of $(CF_3)_2AsCl$ with primary amines in the gas phase.[107] Subsequently, some amino(dichloro)arsanes were described as well.[108-110] Treatment of Ter-NH$_2$ (Ter = 2,6-bis-(2,4,6-trimethylphenyl)) with AsCl$_3$ in the presence of Et$_3$N, *via* aminolysis reaction, resulted in the formation of desired aminoarsane TerN(H)-AsCl$_2$.[111] Later on, the aminolysis reaction between Ar$_2$AsCl and Ar'NH$_2$ led to the formation of aryl derivatives of aminoarsane, such as (p-BrC$_6$H$_4$)$_2$As-N(H)Ph.[112] To the best of our knowledge, metal complexes of arsinoamides have never been reported to date.

2 Aim of the project

Phosphinoamides, which can act as both monodentate and bidentate ligands, have been widely employed to establish main group, transition and lanthanide metal complexes. Bearing the intriguing results of the phosphinoamide chemistry in mind, the heavier arsinoamide congeners ([R$_2$As-NR']$^-$) and their application in coordination chemistry awaits exploration.

Considering the close energy level of atomic orbital, size of atom, electronegativity and polarizability between phosphorus and arsenic,[113-116] it is worthwhile to investigate arsinoamides and their metal complexes. In general, the main target of this thesis is to investigate the synthesis and structural properties of novel arsinoamide ligands and their application in coordination chemistry. In detail, the steps and aims of this project are listed below:

1, synthesis of aminoarsane (R$_2$As-N(H)R') and its deprotonated derivatives (arsinoamide). The aim is to investigate the feature of arsinoamides and the stability of the As-N bond.

2, synthesis of metal complexes ligated by arsinoamides. The aim is to understand the reactivity and coordination mode of arsinoamides towards various metals, such as p-block and d-block metals. The reactivity of arsinoamides will be closely compared to that of the well-established phosphinoamides.

3 Results and discussion

3.1 Synthesis and structural characterization of aminoarsane and alkali metal complexes of arsinoamide (1-4)

This section has already been published in:

Xiao Chen, Michael T. Gamer, and Peter W. Roesky. Synthesis and structural characterization of alkali metal arsinoamides. *Dalton Trans.*, **2018**, 47, 12521-12525.

Schemes 3.1.1-3.1.4 and figures 3.1.1-3.1.5 adapted from Ref. [135] with permission from The Royal Society of Chemistry.

Given their ability to stabilize the partial negative charge on the phosphorus atom by delocalization and thus stabilize the P-N bond, aryl groups, especially phenyl groups, are one of the most common substituents on phosphinoamides. In the case of arsenic compounds, the toxicity should be considered at first. The highly toxic diphenylchlorarsine (Ph_2AsCl) shows a very low melting point (38-42 °C),[117,118] and was even used as a chemical warfare agent. The crude Ph_2AsCl is a liquid at room temperature. Nevertheless, Mes_2AsCl (Mes = 2,4,6-trimethylbenzene) is a light yellow solid under the same conditions. Thus, Mes_2AsCl is much safer and easier to handle than Ph_2AsCl and used as a precursor in this thesis.

Note:

Considering their toxicity towards human and environmental hazard, arsenic compounds should be handled carefully. One of the suggested methods to handle waste arsenic compounds is by chemical precipitation.[119-122] First, the arsenic derivatives are oxidized from the oxidation state +3 to +5 by treatment with H_2O_2, leading to the formation of the $[AsO_4]^{3-}$ anion. Then, $Fe_2(SO_4)_3$ is added, resulting in the precipitation of insoluble $FeAsO_4$, which can be isolated by filtration.

Mes$_2$AsCl was prepared following the published literature procedure. Treatment of AsCl$_3$ with 2 equiv. of 2-mesitylmagnesium bromide in THF resulted in the target compound Mes$_2$AsCl as a light yellow powder.[123] Inspired by the synthesis of aminophosphines, the aminoarsane Mes$_2$AsN(H)Ph (1) was synthesized, *via* an aminolysis reaction, by treatment of Mes$_2$AsCl with aniline in THF (Scheme 3.1.1). Then, triethylamine (Et$_3$N) was used to trap the eliminated HCl. It should be noticed that the white precipitate of Et$_3$N·HCl that is formed is insoluble in THF. After filtration and washing with cold *n*-pentane, Mes$_2$AsN(H)Ph was isolated as a white solid.

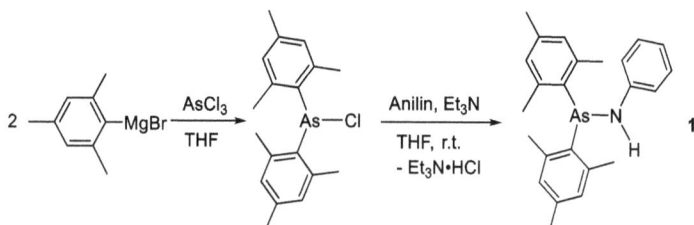

Scheme 3.1.1 Synthesis of Mes$_2$AsN(H)Ph (1).

As a secondary amine, compound 1 features one N-H stretch absorption band v(NH) at 3377 cm^{-1} in the IR spectrum (Figure 3.1.1), which is consistent with the published values.[111,124] In comparison, in the corresponding aminophosphine (Mes$_2$PNHPh), the v(NH) was detected at 3364 cm^{-1}.[125] In the ^{1}H NMR spectrum, the broad signal at 3.98 ppm further proves the presence of the NH group. The *ortho*- and *para*-methyl groups give resonances at 2.35 ppm and 2.06 ppm, respectively. The aromatic signal of the mesityl group is observed at 6.65 ppm. In the ^{13}C{^{1}H} NMR spectrum, the resonances detected at 22.5 ppm and 20.9 ppm correspond to the *ortho*- and *para*-methyl groups of the mesityl ring. The identity of compound 1 was further defined by mass spectrometry with a molecular ion peak at *m/z* 405.

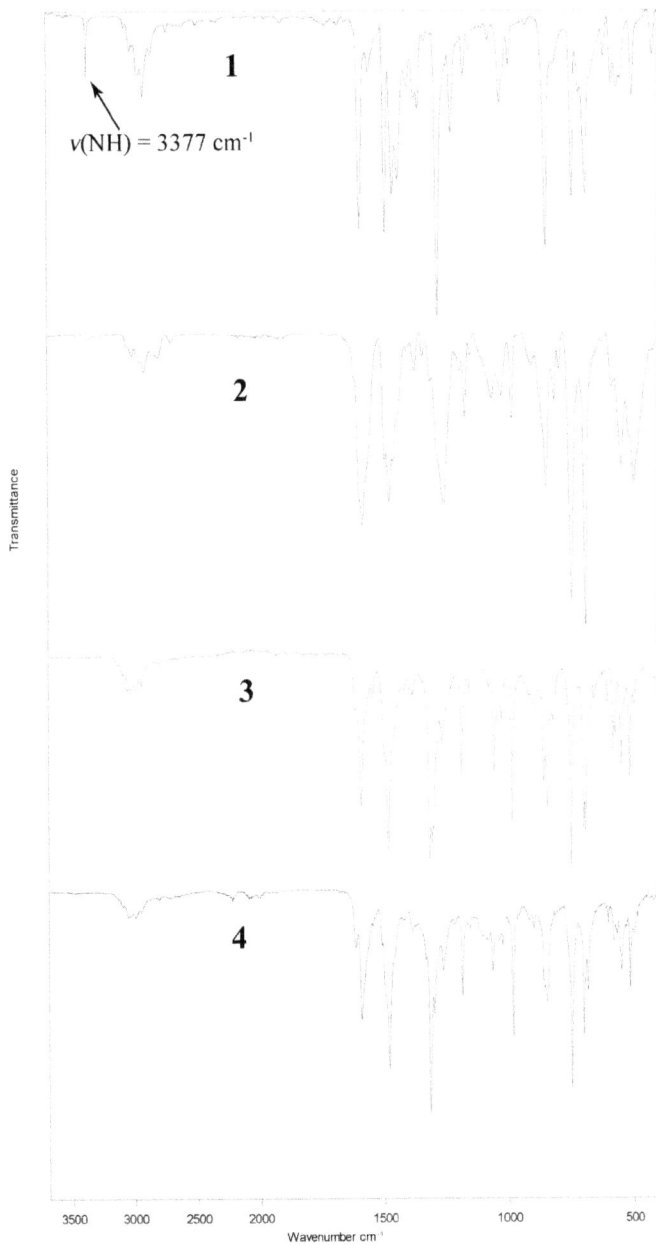

$v(\text{NH}) = 3377 \text{ cm}^{-1}$

Figure 3.1.1 IR spectra of compounds **1-4**.

Deprotonation of **1** with ^{n}BuLi gave the corresponding lithium arsinoamide [(Mes$_2$AsNPh){Li(OEt$_2$)$_2$}] (**2**) (Scheme 3.1.2) as a white solid. In the IR spectrum, no obvious v(N-H) band can be detected from 3500 to 3000 cm^{-1}, which proves the successful deprotonation of compound **1** (Figure 3.1.1). In addition, the NH resonance disappeared in the ^{1}H NMR spectrum of compound **2**, and all of the resonances are slimly down-field shifted in contrast to those of **1**. For example, the resonances of the *ortho*- and *para*-methyl groups are shifted downfield from 2.35 ppm and 2.06 ppm (**1**) to 2.66 ppm and 2.12 ppm (**2**), respectively.

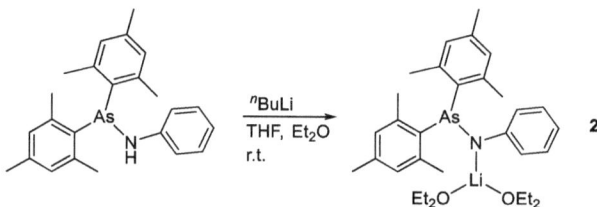

Scheme 3.1.2 Synthesis of [(Mes$_2$AsNPh){Li(OEt$_2$)$_2$}] (**2**).

Colorless crystals of compound **2** were obtained from a mixture solution of *n*-pentane and diethyl ether (1/1) at -30 °C, and the complex crystallized in the monoclinic space group *P*2$_1$ with one molecule in the asymmetric unit. X-ray diffraction studies show that compound **2** shows a monomeric configuration. The lithium atom is three-fold coordinated by one nitrogen atom of the arsinoamide and two oxygen atoms of diethyl ether ligands (Figure 3.1.2). It should be mentioned that one of the diethyl ether positions is about 40% occupied by a THF ligand. However, no Li-As interaction is observed. The As-N bond length is 1.821(7) Å, which fits well with the published values for As-N single bonds (*e.g.* 1.851(2) Å or 1.804(2) Å).[111,126,127] The arsinoamide exhibits a *trans* conformation. In view of the sum of the bonding angles around Li (347.8°), the coordination geometry of the lithium atom can be taken as a distorted trigonal plane.

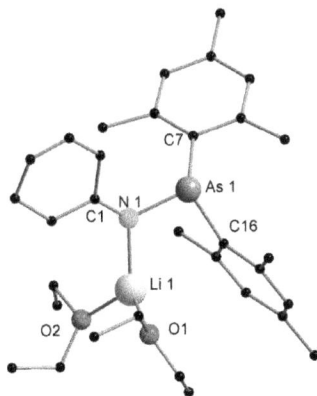

Figure 3.1.2 Molecular structure of compound **2**, [(Mes₂AsNPh){Li(OEt₂)₂}], in the solid state. Hydrogen atoms are omitted for clarity. Selected bond lengths [Å] and angles [°]: As1-N1 1.821(7), As1-C7 1.999(7), As1-C16 1.982(7), Li1-O1 1.93(2), Li1-O2 1.88(2), Li1-N1 1.95(2); As1-N1-Li1 113.6(5), N1-As1-C7 109.2(3), N1-As1-C16 96.3(3).

In contrast, the corresponding phosphinoamide compound [(PhNPPh₂)Li(OEt₂)]₂ forms a dimeric structure, in which the lithium atoms are coordinated by both the nitrogen and phosphorus atoms of phosphinoamide. It should be noticed that the compounds [{(2,6-^{i}Pr₂C₆H₃)NPPh₂}Li(THF)₂] and [{(2,4,6-^{t}Bu₃C₆H₂)NPPh₂}Li(OEt₂)₂], bearing larger substituents on the nitrogen atom, crystallize as monomers instead.[25,128] It is thus anticipated that the steric demand of the substituents plays a crucial role in the aggregation state of the lithium compounds.

After the reaction of compound **1** with NaN(SiMe₃)₂ in THF, the sodium arsinoamide [(Mes₂AsNPh){Na(OEt₂)}]₂ (**3**) was obtained in 76% yield (Scheme 3.1.3). No ν(NH) band or N-H resonance signal was detected in the IR or ^{1}H NMR spectra, respectively (Figure 3.1.1 and 3.1.3), which indicates the successful deprotonation of **1**. The signals at 2.37 ppm and 2.11 ppm are delegated to the *ortho*- and *para*-methyl groups of the mesityl ring, respectively. Owing to the interaction between the aromatic rings and the sodium atom, the aromatic signals (6.66-5.86 ppm) of **3** are broadened and upfield shifted compared to that of compound **1** (7.15-6.65 ppm). This interaction could be further proved by X-ray diffraction analysis.

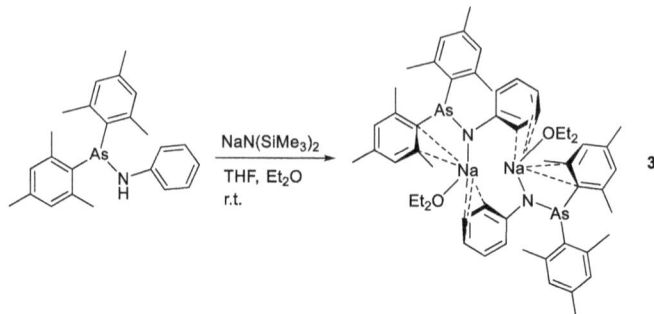

Scheme 3.1.3 Synthesis of [(Mes$_2$AsNPh){Na(OEt$_2$)}]$_2$ (**3**).

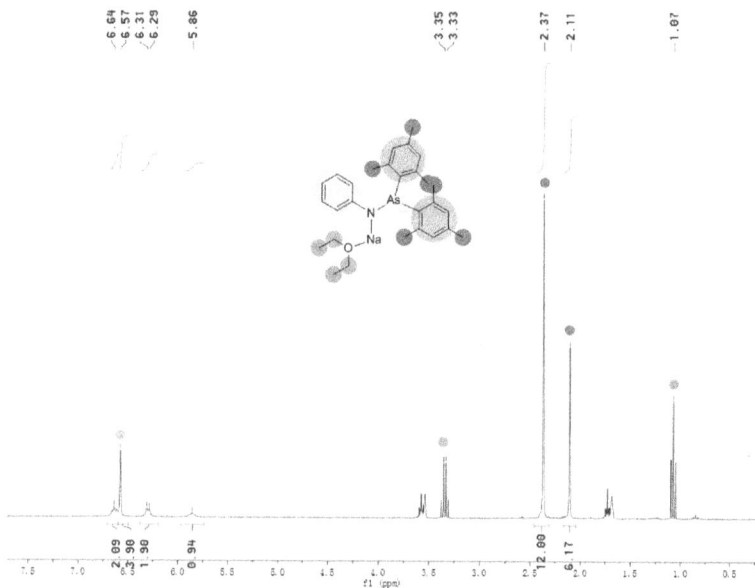

Figure 3.1.3 ^{1}H NMR spectrum of compound **3** in d_8-THF.

Single crystals of compound **3** were grown by layering *n*-heptane onto a solution of **3** in diethyl ether (1/1). The complex crystallized in the triclinic space group *P*-1 (Figure 3.1.4). According to X-ray diffraction studies, compound **3** features a dimeric arrangement with some intermolecular Na-C interactions (2.753(6) Å to 3.040(3) Å).[129] The sodium atom is π-coordinated with one mesityl ring (η^2-fashion) and one phenyl ring (η^3-fashion). In addition, one diethyl ether coordinates to the sodium atom with a Na-O distance of 2.290(4) Å. However, no As-Na interaction can be detected in the molecular structure of **3**. The arsinoamide exhibits a *trans* conformation and the As-N bond length of 1.850(4) Å is close to that of 1.821(7) Å in compound **2**. To the best of our knowledge, no comparable sodium phosphinoamide has ever been reported.

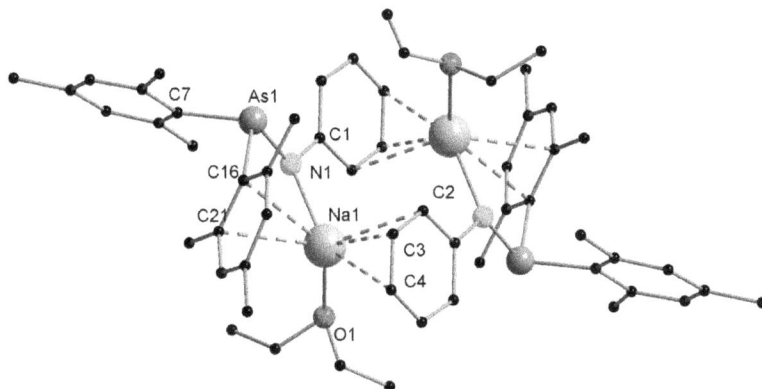

Figure 3.1.4 Molecular structure of compound **3**, [(Mes$_2$AsNPh){Na(OEt$_2$)}]$_2$, in the solid state. Hydrogen atoms are omitted for clarity. Selected bond lengths [Å] and angles [°]: As1-N1 1.850(4), As1-C7 2.018(4), As1-C16 1.996(4), Na1-N1 2.350(4), Na1-C2 2.831(5), Na1-C3 2.753(6), Na1-C4 3.040(3), Na1-C16 2.996(3), Na1-C21 2.944(3), Na1-O1 2.290(5); As1-N1-Na1 113.7(2), Na1-As1-N1 37.62(12), O1-Na1-N1 107.8 (2), N1-As1-C7 109.2(2), C7-As1-Na1 122.69(13), C1-N1-As1 116.5(3), C1-N1-Na1 119.9(3).

The potassium analogue, compound **4** ([(Mes$_2$AsNPh){K(THF)}]$_2$), was obtained from the reaction of compound **1** with KH in THF (Scheme 3.1.4). The ^{1}H NMR spectrum of compound **4** is comparable to that of **3**. For example, the resonances of the *o*- and *p*-methyl groups are detected at 2.39 ppm (2.37 ppm for **3**) and 2.12 ppm (2.11 ppm for **3**), respectively. The interaction between the aromatic rings and the potassium atom is evidenced by broadened and upfield shifted resonances for the aromatic protons (7.15-6.65 ppm in **1** *vs* 6.67-5.83 ppm in **3**). Interaction between the potassium cation and the aryl rings was further supported by X-ray diffraction analyses. In the ^{13}C{^{1}H} NMR spectrum, the resonances at 22.2 ppm and 21.2 ppm can be assigned to the *ortho*- and *para*-methyl of the mesityl ring, respectively.

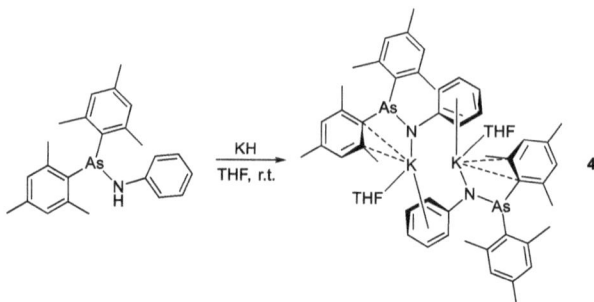

Scheme 3.1.4 Synthesis of [(Mes₂AsNPh){K(THF)}]₂ (4).

Single crystals of compound **4** were grown by layering *n*-heptane onto a solution of **4** in THF, and the complex crystallized in the triclinic space group *P*-1 with one molecule in the unit cell (Figure 3.1.5). According to the X-ray diffraction studies, no K-As interaction can be observed. One THF coordinates to the potassium atom with a K-O distance of 2.608(4) Å.[130] The π-coordinations between K and one phenyl group and one mesityl group in an η^2-fashion and η^6-fashion, respectively were detected, with K-C bond lengths ranging from 3.087(5) Å to 3.291(4) Å.[131] The η^6-fashion coordination causes the dimerization of compound **4**. Compared with the η^3-coordination in **3**, the larger ion radius of the potassium cation can explain the increased coordination number. The arsinoamide exhibits a *trans* conformation and the As-N bond distance (1.843(3) Å) is similar to that in **3** (1.850(4) Å). To the best of our knowledge, no comparable potassium phosphinoamide has ever been reported to date.

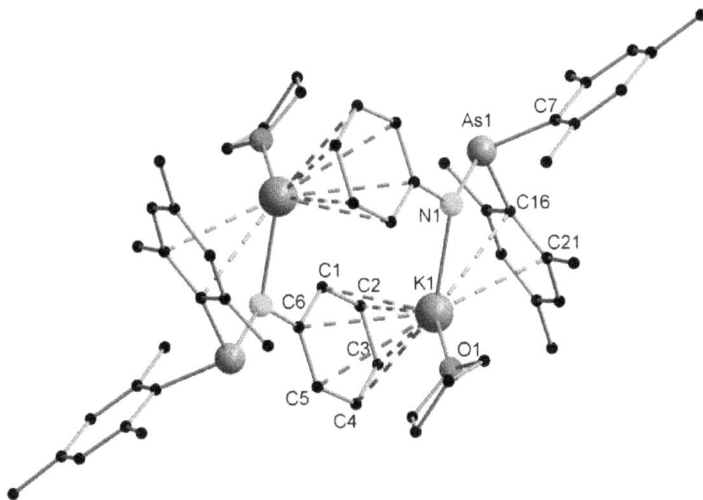

Figure 3.1.5 Molecular structure of compound **4**, [(Mes₂AsNPh){K(THF)}]₂, in the solid state.

Hydrogen atoms are omitted for clarity. Selected bond lengths [Å] and angles [°]: As1-N1

1.843(3), As1-C7 2.010(4), As1-C16 1.984(4), K1-O1 2.608(4), K1-N1 2.749(4), K1-C1

3.110(4), K1-C2 3.087(5), K1-C3 3.168(5), K1-C4 3.256(5), K1-C5 3.291(4), K1-C6 3.238(4),

K1-C16 3.190(4), K1-C21 3.073(4); K1-As1-N1 41.70(11), O1-K1-N1 93.33(11), As1-N1-K1

111.8(2).

Under inert atmosphere, compounds **2**, **3** and **4** are stable in the solid state at room temperature. Nevertheless, in deuterated solvents, a slow decomposition can be detected at room temperature, even in sealed NMR tubes. Furthermore, they rapidly decompose at a temperature of 60 °C, turning into deep brown solutions.

Reaction of anhydrous LnCl₃ (Ln = Sm, Y and Yb) with compound **2** did not lead to the desired homoleptic lanthanide complexes. Attempts to force this reaction by heating the mixture to 80 °C for 3 hours led to the split of the As-N bond and the establishment of Mes₂AsAsMes₂. Thus, a supporting ligand, 2,2'-bipyridine (2,2'-bpy), was introduced in this reaction. Specifically, the reaction of YbCl₃ with 3 equiv. of compound **2** and 1 equiv. of 2,2'-bpy was considered. Unfortunately, single crystals of compound **5** ([(Mes₂AsNPh){Li(2,2'-bpy)}]) were exclusively obtained from toluene upon cooling to -30 °C (Figure 3.1.6).

Compound **5** crystallized in the triclinic space group *P*-1 with four molecules of **5** and four molecules of toluene in the unit cell. After the leaving of two diethyl ether molecules, the 2,2'-bpy ligand chelates the lithium atom, leading to the formation of two N-Li bonds with distances of 2.009(7) Å and 2.078(7) Å. Comparable Li-N bond distances could be found in the published literature.[132,133] Additionally, X-ray diffraction studies revealed two Li-C interactions (2.631(8) Å and 2.607(9) Å) and the presence of one free toluene molecule in the structure. One of the most remarkable features is the formation of an Li-As interaction with a distance of 3.085(7) Å, which is *ca.* 0.08 Å shorter than the longest reported Li-As interaction of 3.162(10) Å.[134] It seems that the formation of the Li-As interaction is due to a variation in the electron density on the lithium atom. Selected bond distances and angles of compounds **2** and **5** are listed in Table 3.1.1.

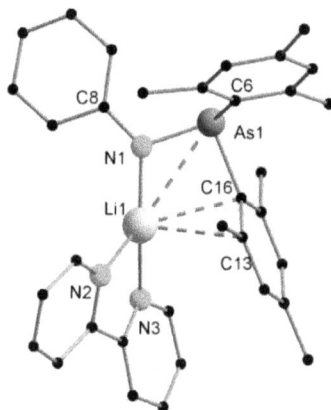

Figure 3.1.6 Molecular structure of compound **5**, [(Mes$_2$AsNPh){Li(2,2'-bpy)}], in the solid state. Toluene and hydrogen atoms are omitted for clarity. Selected bond lengths [Å] and angles [°]: As1-N1 1.839(3), As1-C6 1.998(3), As1-C16 1.982(3), As1-Li1 3.085(7), N1-C8 1.385(5)· N1-Li1 1.943(7), N3-Li1 2.078(7), N4-Li1 2.009(7), C16-Li1 2.631(8), C13-Li1 2.607(9); N1-As1-C6 111.1(2), N1-As1-C16 94.4(2), N1-As1-Li1 36.4(2), C6-As1-Li1 116.7(2), C16-As1-C6 101.0(2), C16-As1-Li1 58.0(2), As1-N1-Li1 109.2(3), C8-N1-As1 115.6(2), C8-N1-Li1 129.4(3), As1-C16-Li1 82.2(2).

Table 3.1.1 Bond lengths and angles of compounds **2** and **5**.

	[(Mes$_2$AsNPh){Li(OEt$_2$)$_2$}] (2)	[(Mes$_2$AsNPh){Li(2,2'-bpy)}] (5)
As-N bond length (Å)	1.816(6)	1.839(3)
Li-N bond length (Å)	1.951(13)	1.932(7)
Li-As bond length (Å)	-	3.085(7)
Li-N-As angle (°)	113.6(5)	109.2(3)

In summary, the aminoarsane Mes$_2$AsN(H)Ph (**1**) and its alkali metal derivatives (**2** = [(Mes$_2$AsNPh){Li(OEt$_2$)$_2$}], **3** = [(Mes$_2$AsNPh){Na(OEt$_2$)}]$_2$, **4** = [(Mes$_2$AsNPh){K(THF)}]$_2$) have been prepared and fully characterized. According to X-ray diffraction studies, compound **2** forms a monomer, while **3** and **4** feature dimer arrangements with some M-C interactions between the metal atom and the aromatic ring. Unfortunately, no M-As interaction can be detected in the alkali metal etherate derivatives.

3.2 Synthesis and structural characterization of group 4 metal complexes of arsinoamide (6-9)

As mentioned above, the attempted synthesis of homoleptic lanthanide complexes of arsinoamides failed. Considering the well-established coordination chemistry of phosphinoamides with zirconium, we moved to the neighbours of lanthanides, which are group 4 elements, and attempted to synthesize the corresponding arsinoamide metal complexes.

Inspired by the synthetic route to phosphinoamide zirconium complexes, a salt metathesis procedure should be the most convenient approach to form arsinoamide zirconium complexes. Transmetallation of $ZrCl_4$ with 2 equiv. of compound **2** in THF resulted in the corresponding bis-substituted complex **6** ($[(Mes_2AsNPh)_2ZrCl_2(THF)]\cdot(DCM)$) after recrystallization from DCM (Scheme 3.2.1). Compound **6** was fully characterized by infrared spectroscopy, elemental analysis and nuclear magnetic resonance, and the molecular structure was established by X-ray diffraction analysis.

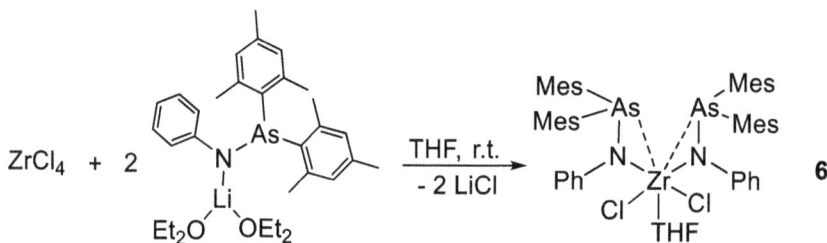

Scheme 3.2.1 Synthesis of $[(Mes_2AsNPh)_2ZrCl_2(THF)]$ (**6**).

It is notable that the steric effects of the phosphinoamide ligand can significantly influence the nature of the products in the synthesis of phosphinoamide zirconium complexes.[59,66,68] For example, after the reaction of $ZrCl_4$ with 2 equiv. of $[PhN(Li)PPh_2]$, the tris(phosphinoamide)zirconium complex is isolated in general, instead of the bis-substituted complex. Due to the bulk of the mesityl ring, only the bis(arsinoamide)zirconium complex **6** was isolated successfully after the reaction of $ZrCl_4$ with 2 equiv. of compound **2**. In the ^{1}H NMR spectrum, compared with that of **2**, a slight upfield shift could be detected for the o- and p-methyl groups of the mesityl ring (2.12 ppm and 1.69 ppm in **6** vs 2.66 ppm and 2.12 ppm in

2).[135] The singlet at 6.50 ppm belongs to the aromatic signal of the mesityl group. In the $^{13}C\{^1H\}$ NMR spectrum, the *ortho*- and *para*-methyl groups give resonances at 22.6 ppm and 20.7 ppm, respectively.

Compound **6** crystallized by slow evaporation of a dichloromethane (DCM) solution of the complex in the monoclinic space group $P2_1/n$ (Figure 3.2.1). Single crystal X-ray diffraction analyses indicate the existence of two Zr···As interactions with distances of 3.2105(9) Å and 3.2201(9) Å, which are *ca.* 0.2 Å greater than the published As-Zr interaction values.[134,136,137] Additionally, given the sum of the Pauling's covalent radii of Zr and As (1.75(Zr) + 1.19(As) = 2.94 Å),[138,139] the Zr···As interaction in **6** should be very weak. The Zr-N bond lengths (2.041(4) Å and 2.037(4) Å) are consistent with the published values.[66,140-142] In addition, the presence of a Zr···C interaction and the coordination of one molecule of THF to the zirconium atom are observed with bond lengths of 2.832(5) Å (Zr···C) and 2.207(3) Å (Zr-O), respectively.[143,144] The N1-Zr-N2 angle of 118.5(2)° is consistent with that of 113.66(12)° in $[\{(Me_3Si)_2N\}_2ZrCl_2(THF)]$.[145] According to the molecular structure, the arsinoamides adopt a *cis* conformation.

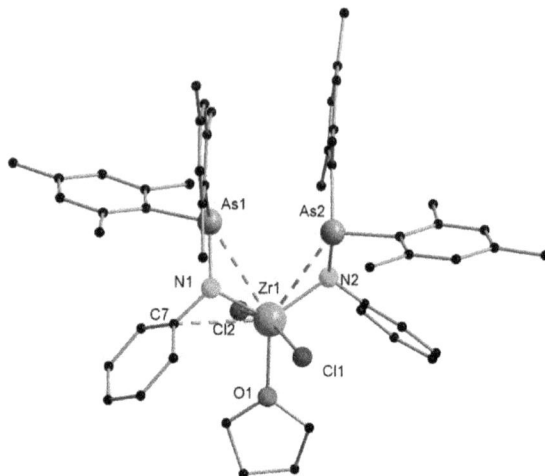

Figure 3.2.1 Molecular structure of compound **6**, [(Mes₂AsNPh)₂ZrCl₂(THF)]·(DCM), in the solid state. Non-coordinating solvent molecules (DCM) and hydrogen atoms have been omitted for clarity. Selected bond lengths [Å] and angles [°]: Zr1-As1 3.2105(9), Zr1-As2 3.2201(9), Zr1-Cl1 2.4840(13), Zr1-Cl2 2.4930(13), Zr1-N1 2.041(4), Zr1-N2 2.037(4), As1-N1 1.869(3), Zr1-C7 2.832(5), As2-N2 1.887(4) Zr1-O1 2.207(3); As1-Zr1-As2 68.06(3), N1-Zr1-As1 33.10(10), N1-Zr1-As2 90.10(10), N2-Zr1-N1 118.5(2), As1-N1-Zr1 110.3(2), C7-N1-As1 131.4(3).

Compound **7** ([(Mes₂AsNPh)₂HfCl₂(THF)]·(DCM)) was synthesized following the same synthetic approach as that carried out for **6** but using HfCl₄ instead of ZrCl₄. After the reaction of HfCl₄ with 2 equiv. of compound **2** in THF, *via* a salt metathesis reaction, compound **7** was isolated as a white solid with yield of 47 % (Scheme 3.2.2). The NMR spectra of **7** are similar to that of compound **6**. In the ¹H NMR spectrum, the *ortho*- and *para*-methyl groups of the mesityl ring give resonances at 2.10 ppm and 1.69 ppm, respectively. The resonance at 6.50 ppm is assigned to the aromatic signal of the mesityl group. In the ¹³C{¹H} NMR spectrum, the methyl groups of the mesityl ring give resonances at 22.8 ppm (*o*-CH₃) and 20.7 ppm (*p*-CH₃).

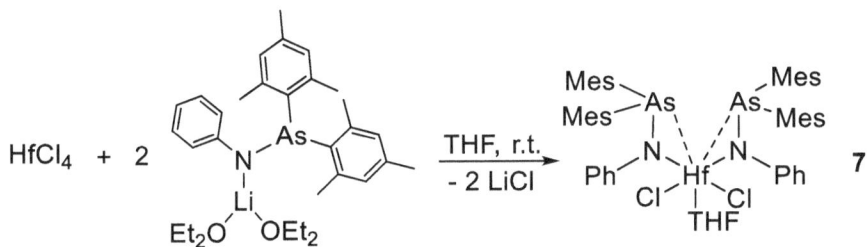

Scheme 3.2.2 Synthesis of [(Mes₂AsNPh)₂HfCl₂(THF)] (7).

Colorless crystals of compound **7** were isolated from DCM by slow evaporation, and the complex crystallized in the monoclinic space group $P2_1/n$ (Figure 3.2.2). X-ray diffraction analyses indicate that compounds **7** and **6** are isostructural, probably due to the comparable ionic radius of Zr and Hf. Two weak Hf···As interactions could be detected with distances of 3.2442(4) Å and 3.2343(4) Å, which are similar to the distance of the Zr···As interaction in **6**. It should be noted that only two structurally characterized Hf-As bonds have been reported to date. The first Hf-As bond, with a distance of 2.7257(11) Å, was observed in [Cp'₂HfCl{As(SiMe₃)₂}].[146] Webster described the other complex, [HfI₄{o-C₆H₄(AsMe₂)₂}₂], with Hf-As bond lengths varying from 2.8665(7) to 2.8966(7) Å.[136] The longer Hf-As bond distances in compound **7** may relate to the bulkiness of the arsinoamide ligands. The Hf-N bond lengths (2.016(3) Å and 2.024(3) Å) are in good agreement with the reported single-bond values.[147,148] Same as in compound **6**, one Hf···C interaction was observed with a distance of 2.798(3) Å. Additionally, the N-M-N bond angle (N1-Hf-N2: 115.59(12)°) is similar to that in **6** (N1-Zr-N2: 118.5(2)°).

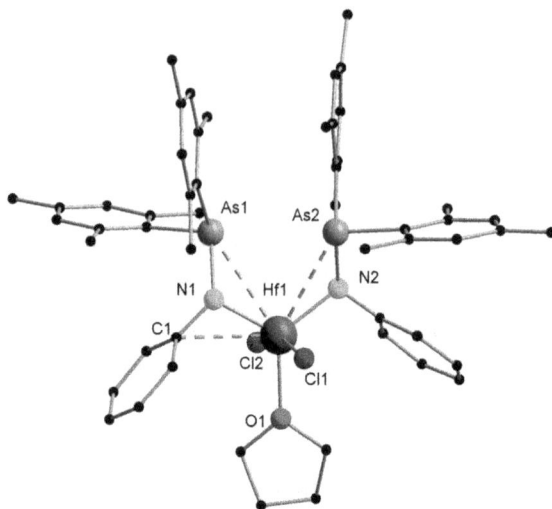

Figure 3.2.2 Molecular structure of compound **7**, [(Mes$_2$AsNPh)$_2$HfCl$_2$(THF)]·(DCM), in the solid state. Non-coordinating solvent molecules (DCM) and hydrogen atoms have been omitted for clarity. Selected bond lengths [Å] and angles [°]: Hf1-As1 3.2442(4), Hf1-As2 3.2343(4), Hf1-Cl1 2.4540(9), Hf1-Cl2 2.4657(10), Hf1-O1 2.179(3), Hf1-N1 2.016(3), Hf1-N2 2.024(3), Hf1-C1 2.798(3), As1-N1 1.874(3), As2-N2 1.882(3); As2-Hf1-As1 66.816(10), N1-Hf1-As1 32.12(9), N1-Hf1-N2 115.59(12), N2-Hf1-As1 86.75(8), N2-Hf1-As2 32.70(9), C1-N1 As1 129.9(2).

In addition, the reaction of TiCl$_4$ with 2 equiv. of compound **2** was also considered. Unfortunately, decomposition of compound **2** occurred and only crystals of the by-product, Mes$_2$AsAsMes$_2$, were isolated. In contrast with the well-established titanium phosphinoamides complexes, it is assumed that the As-N bond is seemingly not as strong as the P-N bond.

Considering the weak M···As interactions in compounds **6** and **7**, we aimed to have a thorough study of this interaction by changing the nature of the substituents on the metal atom. Specifically, to further investigate the substituent effect on the M···As interaction, the reactions between the amido complexes [(Me$_2$N)$_2$MCl$_2$(THF)$_2$] (M = Zr, Hf) and compound **2** were considered.

Treatment of [(Me$_2$N)$_2$ZrCl$_2$(THF)$_2$] with 2 equiv. of compound **2** in THF, *via* a salt metathesis reaction, resulted in the corresponding bis-substituted product **8** ([(Mes$_2$AsNPh)$_2$Zr(NMe$_2$)$_2$]) (Scheme 3.2.3). In the ^{1}H NMR spectrum (Figure 3.2.3), the singlets at 2.32 ppm and 2.07 ppm belong to the *ortho*- and *para*-methyl groups of the mesityl ring, respectively. Compared with a signal at 2.63 ppm for compound **2**, a slight upfield shift of the *ortho*-methyl groups (2.32 ppm in **8**) could be seen. In addition, the resonance of the two equivalent dimethylamide groups is detected at 2.95 ppm, which is consistent with that at 3.31 ppm for [(Me$_2$N)$_2$ZrCl$_2$(THF)$_2$].[149] In the ^{13}C{^{1}H} NMR spectrum, the characteristic resonance of the dimethylamide groups is observed at 44.0 ppm. The singlets at 22.9 ppm and 21.0 ppm are assigned to the *ortho*- and *para*-methyl groups, respectively.

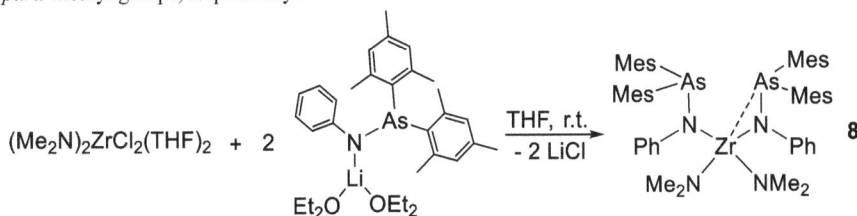

Scheme 3.2.3 Synthesis of [(Mes$_2$AsNPh)$_2$Zr(NMe$_2$)$_2$] (**8**).

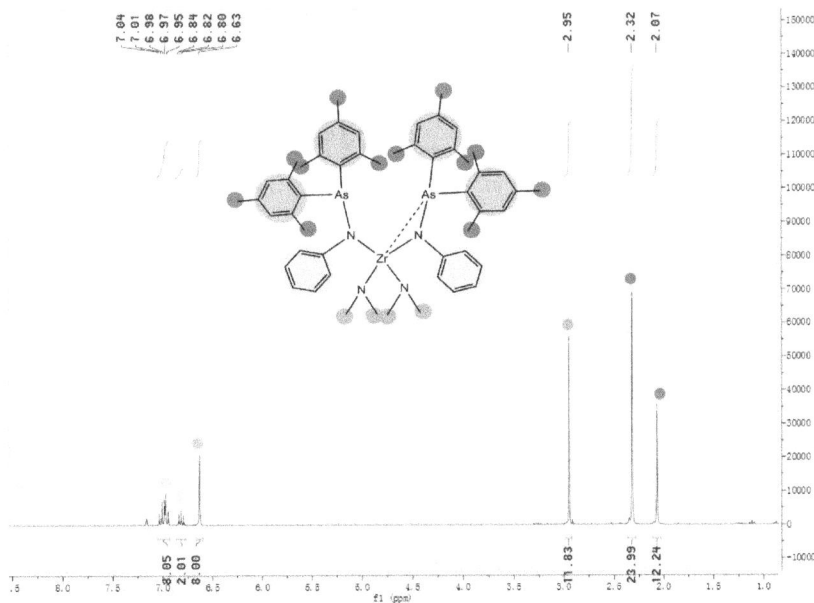

Figure 3.2.3 ^{1}H NMR spectrum of compound **8** in C$_6$D$_6$.

Colorless crystals of compound **8** were obtained from a mixture of *n*-pentane and diethyl ether (1/1) at -30 °C, and the complex crystallized in the triclinic space group *P*-1 with two molecules in the unit cell (Figure 3.2.4). Compared with Zr···As separations of 3.2105(9) Å and 3.2201(9) Å in compound **6**, a shorter Zr···As interaction (3.0770(4) Å) can be observed in **8**. Still, this distance is slightly longer than the reported values (longest described value of 2.9979 Å).[150-152] However, only one Zr···As interaction is observed, which may be attributed to the change in the electron density and coordination environment at the zirconium atom. Due to the steric effect of the arsinoamides, the Zr1-N1 and Zr1-N2 bond lengths are slimly longer than the Zr1-N3 and Zr1-N4 bonds with the dimethylamide ligands. Additionally, the N1-Zr1-N2 bond angle (118.86(11)°) is greater than N3-Zr1-N4 (103.40(13)°).[153] In general, the zirconium atom is five-fold coordinated, giving a slightly distorted trigonal bipyramidal geometry. As observed for compound **6**, the geometry of the nitrogen atoms (N1 and N2) in compound **8** could be best described as a trigonal planar. However, the sums of the angles are slightly greater than those in compound **6** (350.99° and 350.73° in **6** *vs* 356.55° and 356.38° in **8**), which may be associated with the different steric demand of the substituents on the zirconium atom. According to the molecular structure, the arsinoamide ligands exhibit a *cis* conformation.

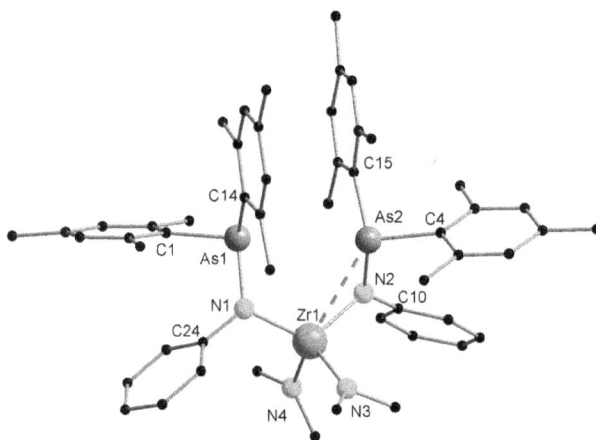

Figure 3.2.4 Molecular structure of compound **8**, [(Mes$_2$AsNPh)$_2$Zr(NMe$_2$)$_2$], in the solid state. Hydrogen atoms have been omitted for clarity. Selected bond lengths [Å] and angles [°]: Zr1-As2 3.0770(5), Zr1-N1 2.083(3), Zr1-N2 2.097(3), Zr1-N3 2.042(3), Zr1-N4 2.036(3), As1-N1 1.878(3), As2-N2 1.861(3), As1-C1 1.966(3), As1-C14 1.986(3), As2-C4 1.966(3), As2-C15 1.977(3); N2-Zr1-As2 36.28(8), N3-Zr1-N2 101.56(12), N4-Zr1-N3 103.40(13), N1-Zr1-N2 118.86(11), As2-N2-Zr1 101.88(12), As1-N1-Zr1 114.56(13).

Compound **9** ([(Mes$_2$AsNPh)$_2$Hf(NMe$_2$)$_2$]) was obtained following the same synthetic approach as carried out for **8** but using [(Me$_2$N)$_2$HfCl$_2$(THF)$_2$] instead of [(Me$_2$N)$_2$ZrCl$_2$(THF)$_2$]. After the reaction of [(Me$_2$N)$_2$HfCl$_2$(THF)$_2$] with 2 equiv. of compound **2** in tetrahydrofuran, compound **9** was isolated as a white solid (Scheme 3.2.4). A one-pot reaction was considered as well. After the reaction of ZrCl$_4$ with 2 equiv. of compound **2** in THF for 4 hours, 2 equiv. of LiNMe$_2$ were added into the mixture and the reaction was stirred for further 12 h. Unfortunately, we failed to isolate the pure product **9** from this procedure.

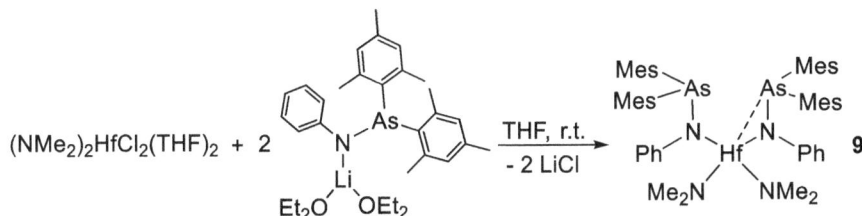

Scheme 3.2.4 Synthesis of [(Mes$_2$AsNPh)$_2$Hf(NMe$_2$)$_2$] (**9**).

In the ^{1}H NMR spectrum, the *ortho-* and *para*-methyl of the mesityl groups give resonances at 2.31 ppm and 2.07 ppm, respectively. The dimethylamide groups exhibit a single resonance at 2.97 ppm, which agrees with that at 2.39 ppm for [(Me$_2$N)$_2$HfCl$_2$(THF)$_2$].[149] In the ^{13}C{^{1}H} NMR spectrum, the resonances at 22.9 ppm and 20.9 ppm are assigned to the *p-* and *o*-methyl of the mesityl groups, respectively. The dimethylamide groups give rise to a signal at 43.5 ppm.

Colorless crystals of compound **9** were obtained from a mixture solution of *n*-pentane and diethyl ether (1/1) at -30 °C, and the complex crystallized in the triclinic space group *P*-1 with four molecules in the unit cell (Figure 3.2.5). Single crystal X-ray diffraction studies indicate that compounds **9** and **8** are isostructural, probably owing to the comparable ionic radii of Zr and Hf. Compared with compound **7**, only one weak Hf···As interaction is detected with a shorter separation (3.0894(4) Å), which is *ca.* 0.2 Å longer than the longest reported Hf-As bond distance (2.8966(7) Å).[136] Due to the bulkiness of the arsinoamides, the Hf1-N1 and Hf1-N2 bonds are slightly longer than the Hf1-N3 and Hf1-N4 bonds. In addition, the N1-Hf1-N2 bond angle (118.90(9)°) is greater than N3-Hf1-N4 (103.57(11)°). The sums of the bonding angles around the nitrogen atoms are greater than those in **7** (356.97° and 356.83° in **9** *vs* 350.03° and

350.75° in **7**). The hafnium atom is five-fold coordinated, resulting in a slightly distorted trigonal bipyramidal coordination geometry.

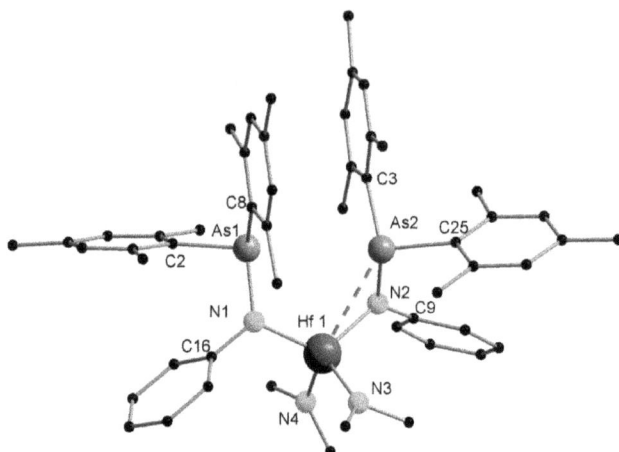

Figure 3.2.5 Molecular structure of compound **9**, [(Mes$_2$AsNPh)$_2$Hf(NMe$_2$)$_2$], in the solid state. Hydrogen atoms have been omitted for clarity. Selected bond lengths [Å] and angles [°]: Hf1-As2 3.0894(4), Hf1-N1 2.067(3), Hf1-N3 2.030(4), Hf1-N2 2.079(3), Hf1-N4 2.036(4), As1-N1 1.888(3), As2-N2 1.872(3), As1-C2 1.977(4), As1-C8 1.987(4); N3-Hf1-N4 103.4(2), N1-Hf1-N2 118.93(12), As1-N1-Hf1 114.9(2), As2-N2-Hf1 102.7(2), N2-Hf1-As2 36.24(9), N3-Hf1-N2 107.45(13), N1-Hf1-As2 92.89(8).

In compounds **6-9**, one of the most remarkable differences lies in the distance of the M···As interaction. In compounds **6** and **7**, the distances of the M···As interaction range from 3.2105(9) Å to 3.2442(4) Å. However, much shorter interactions (3.0770(5) Å (**8**) and 3.0894(4) Å (**9**)) are observed in compounds **8** and **9**. It is suggested that the M···As bond length is notably affected by the coordination environment on the metal atom. A similar trend can be found in related diphosphinoamides zirconium complexes, where the Zr-P interaction could be notably affected by the nature of the substituents on the zirconium atom (*e.g.* 2.6565(5) Å (chloride substituent) *vs* 2.7208(8) Å (benzyl substituent)).[153] Selected bond lengths of compounds **6-9** are listed in Table 3.2.1. In summary, the arsinoamides complexes **6-9** provide another pathway to investigate the formation and properties of heterobimetallic early/late transition metal complexes.

Table 3.2.1 Distances of M···As interaction in compounds **6-9**.

Compound	M···As interactions (Å)
[{Mes$_2$AsNPh}$_2$ZrCl$_2$(THF)]·(DCM) (**6**)	3.2105(9), 3.2201(9)
[{Mes$_2$AsNPh}$_2$HfCl$_2$(THF)]·(DCM) (**7**)	3.2442(4), 3.2343(4)
[{Mes$_2$AsNPh}$_2$Zr(NMe$_2$)$_2$] (**8**)	3.0770(5)
[{Mes$_2$AsNPh}$_2$Hf(NMe$_2$)$_2$] (**9**)	3.0894(4)

To further understand the substituent effect on the M···As interaction, subsequent reaction of compound **6** with 2 equiv. of MeMgBr was considered, with the aim to synthesize the dialkyl product [(Mes$_2$AsNPh)$_2$Zr(CH$_3$)$_2$]. Unfortunately, only crystals of the [(Mes$_2$AsNPh)MgBr(THF)]$_2$ (**10**) by-product were isolated (Figure 3.2.6), resulting from a ligand rearrangement between the zirconium and magnesium centers. The complex crystallized in the monoclinic space group $P2_1/n$ with two molecules in the unit cell. XRD analyses reveal that **10** consists of a bromide-bridged dimer with a Mg$_2$Br$_2$ core[154,155] and with the arsinoamides ligands arranged in a *trans* fashion across the ring. The Mg-N bond length (1.981(4) Å) is consistent with the published values.[156-158] The magnesium atom is four-fold coordinated by one nitrogen atom of the arsinoamide, one oxygen atom of THF and two bromine atoms, leading to a distorted tetrahedral coordination geometry. The arsinoamides adopt a *trans* conformation. However, no Mg-As interaction is observed. It should be noted that Mg-P interactions (2.6318(7) Å to 2.9212(7) Å) have been observed in the magnesium phosphinoamide complex [(Ph$_2$PNDipp)$_2$Mg]$_2$.[39] This discrepancy may relate to the different steric demand of the ligands or the energetically less favored Mg···As interaction.

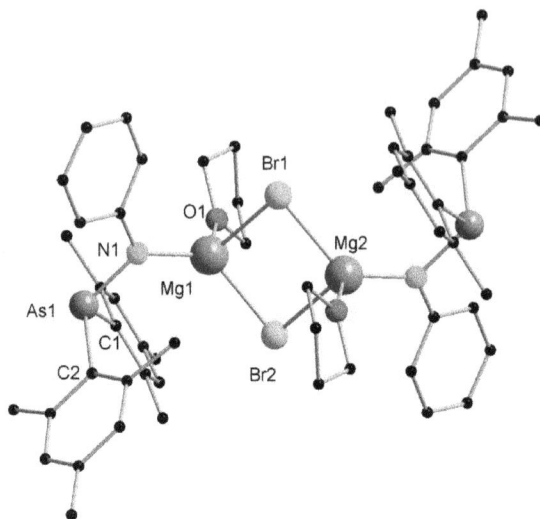

Figure 3.2.6 Molecular structure of compound **10**, [(Mes$_2$AsNPh)MgBr(THF)]$_2$, in the solid state. Hydrogen atoms have been omitted for clarity. Selected bond lengths [Å] and angles [°]: As1-N1 1.851(5), As1-C1 1.987(4), As1-C2 1.977(5), Mg1-N1 1.981(4), Mg1-O1 1.983(4), N1-As1-C1 95.99(8), C1-As1-C1 105.58(7), Mg1-Br1-Mg2 88.01(6), Br1-Mg1-Br2 91.99(6), N1-Mg1-Br2 111.06(8), N1-Mg1-Br1 129.43(7), N1-Mg1-O1 118.46(8), As1-N1-Mg1 120.1(2).

In a further investigation of the coordination behavior of arsinoamide ligands with early transition metals, the reactions of compound **2** with TaCl$_5$ and NbCl$_5$ were considered. Unfortunately, they all led to the decomposition of compound **2**. The reaction of TaCl$_5$/NbCl$_5$ with 2 equiv. of compound **2** in tetrahydrofuran led in the exclusive isolation of Mes$_2$AsAsMes$_2$ as a side-product. Furthermore, under the same conditions, using 3 equiv. of compound **2**, only crystals of compound **11** ([(Mes$_2$As)$_2$NPh]) were isolated (Figure 3.2.7). According to X-ray diffraction analyses, the nitrogen atom bonds with two AsMes$_2$ groups, leading to the formation of an As-N-As unit. The As-N bond lengths (1.899(7) Å and 1.894(6) Å) are slightly longer than that of 1.821(7) Å in compound **2**. The As1-N1-As2 angle (102.5(3)°) is smaller than that of 120° in [(Me$_2$As)$_3$N] and [(MeC$_6$H$_4$S$_2$As)$_3$N]. However, the nitrogen atom exhibits a perfect trigonal planar geometry (sum of the bonding angles around the nitrogen atom: 360.0°).

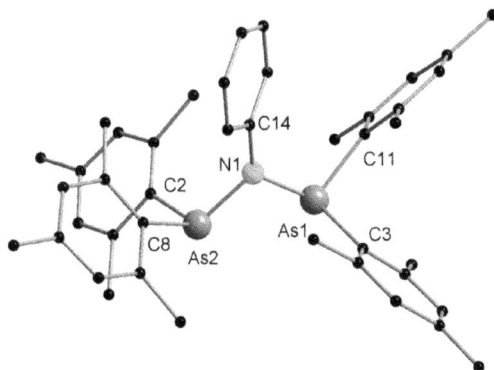

Figure 3.2.7 Molecular structure of compound **11**, [(Mes₂As)₂NPh], in the solid state. Hydrogen atoms have been omitted for clarity. Selected bond lengths [Å] and angles [°]:As1-N1 1.901(7), As2-N1 1.894(6), N1-C14 1.433(11); As1-N1-As2 102.5(3), N1-As1-C2 111.5(4), N1-As1-C8 105.4(3), C14-N1-As1 128.5(5), C14-N1-As2 129.1(5).

It should be mentioned that a similar decomposition was reported in the preparation of phosphinoamide-supported niobium complexes. During the synthesis of the [*i*PrN-PPh₂]⁻ substituted Nb(V) complexes, cleavage of the P-N bond was detected, resulting in the formation the Nb(V) imido complex (Nb=N*i*Pr).[159,160]

3.3 Synthesis and structural characterization of group 13 metal complexes of arsinoamide (12 and 13)

Similar to zirconium and hafnium, group 13 elements are usually classified as Lewis acids. Additionally, Power suggested the presence of an interaction between the lone pair of Bi and the vacant orbital of group 13 elements.[161-163] Considering the above reasons, we moved to the synthesis of arsinoamide complexes with group 13 compounds.

After the reaction of $AlCl_3$ with 1 equiv. of compound **2** in THF, *via* salt metathesis, compound **12** [(Mes$_2$AsNPh)AlCl$_2$(THF)] was isolated as a white solid (Scheme 3.3.1). Interestingly, after the mixing of $AlCl_3$ and compound **2**, the reaction mixture turned gray. In contrast, the THF solution of the mixture was light yellow, which may be related to the coordination of THF to the Al center.

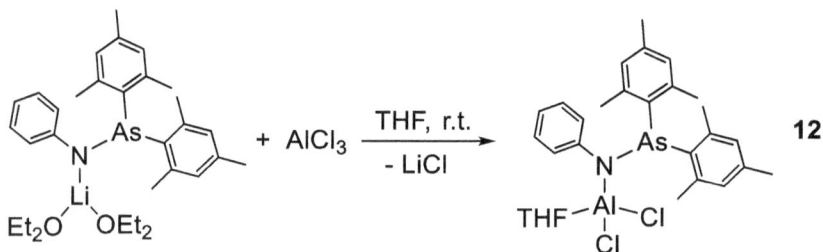

Scheme 3.3.1 Synthesis of [(Mes$_2$AsNPh)AlCl$_2$(THF)] (**12**).

The NMR spectra of compound **12** are similar to those of **2**. In the ^{1}H NMR spectrum, the characteristic resonances of the *ortho*- and *para*-methyl groups in the mesityl ring are seen at a similar chemical shift range (2.62 ppm and 2.11 ppm in **12** *vs* 2.66 ppm and 2.12 ppm in **2**). The singlet at 6.79 ppm is assigned to the aromatic signal of the mesityl group (6.72 ppm in **2**). The resonances at 23.0 ppm and 21.0 ppm belong to the methyl groups of the mesityl ring (22.0 ppm and 21.0 ppm in **2**) in the ^{13}C{^{1}H} NMR spectrum.

Single crystals of compound **12** were obtained by slow evaporation a DCM solution of **12**. The complex crystallized in the triclinic space group *P*-1 (Figure 3.3.1). The Al-N distance (1.823(3) Å) is in the range of the published values for Al-N single bonds.[164-166] One molecule of THF coordinates to the aluminum atom with an Al-O distance of 1.868(3) Å. The As1-N1-Al1 angle

(125.80(14)°) is greater than that of compound **2** (As1-N1-Li1: 113.6(5)°), which can be explained by the nature of the metal atom. The arsinoamide exhibits a *trans* conformation. The aluminum atom is four-fold coordinated by one nitrogen atom, two chloride atoms and one oxygen atom of THF, displaying a distorted tetrahedral geometry. Besides, the nitrogen atom exhibits an almost ideally trigonal planar geometry, in view of the sum of the angles around the nitrogen atom of 357.4°. According to the crystal structure, no As-Al interaction is observed in compound **12**.

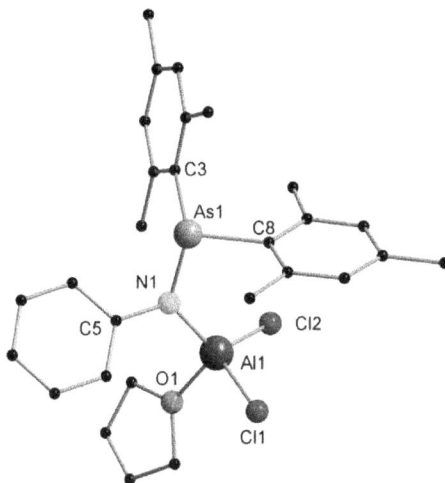

Figure 3.3.1 Molecular structure of compound **12**, [(Mes$_2$AsNPh)AlCl$_2$(THF)], in the solid state. Hydrogen atoms have been omitted for clarity. Selected bond lengths [Å] and angles [°]: As1-N1 1.882(3), As1-C3 1.993(3), As1-C8 1.986(3), Cl1-Al1 2.1406(14), Cl2-Al1 2.119(2), Al1-N1 1.823(3), Al1-O1 1.868(3), N1-C5 1.430(4); N1-As1-C3 110.42(14), N1-As1-C8 99.30(12), Cl2-Al1-Cl1 111.08(6), Al1-N1-As1 125.80(14), C5-N1-As1 111.1(2), C5-N1-Al1 120.5(2), O1-Al1-Cl1 100.72(10), O1-Al1-Cl2 100.97(10), N1-Al1-Cl1 117.51(11), N1-Al1-Cl2 118.41(11).

Reaction of InCl$_3$ with 1 equiv. of compound **2** in THF resulted in the unexpected complex **13** ([(Mes$_2$AsNPh)InCl$_3$][Li(THF)$_4$]), which consists of an ion pair composed of one [{(Mes$_2$AsNPh)InCl$_3$}$^-$] anion and one [{Li(THF)$_4$}$^+$] cation (Scheme 3.3.2). In the ^{1}H NMR spectrum, the characteristic resonances of the *ortho*- (2.36 ppm) and *para*-methyl (2.13 ppm) groups in the mesityl ring are detected. In addition, an upfield shift of the phenyl ring resonances could be observed (6.72-5.94 ppm (**13**) *vs* 7.33-6.63 ppm (**2**)). In the ^{13}C{^{1}H} NMR spectrum, the resonances at 21.7 ppm and 20.7 ppm are assigned to the *ortho*- and *para*-methyl groups of the mesityl group, respectively.

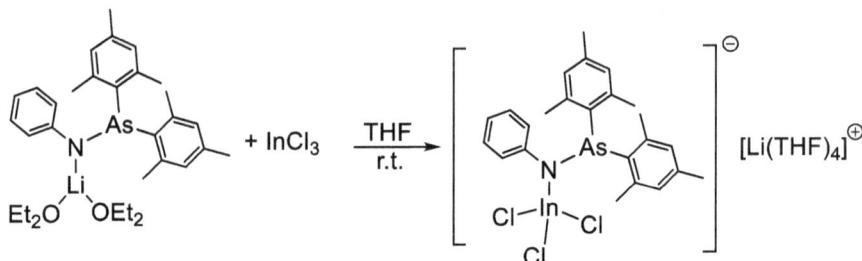

Scheme 3.3.2 Synthesis of [(Mes$_2$AsNPh)InCl$_3$][Li(THF)$_4$] (13).

Single crystals of compound 13 were grown from a mixture of THF and diethyl ether (1/1) stored at -30 °C, and the complex crystallized in the monoclinic space group $P2_1/n$ (Figure 3.3.2). X-ray diffraction analyses confirmed the formation of the unexpected [(Mes$_2$AsNPh)InCl$_3$][Li(THF)$_4$] ion pair. The In-N bond distance (2.075(2) Å) fits well with the published values.[167-169] The As1-N1-In1 bond angle (126.40(11)°) is similar to the As1-N1-Al1 angle (125.80(14)°) in compound 12. The indium atom is four-fold coordinated, revealing a distorted tetrahedral geometry. According to the crystal structure, no As-In interaction is observed in compound 13. To the best of our knowledge, no comparable indium phosphinoamide complex has ever been structurally characterized.

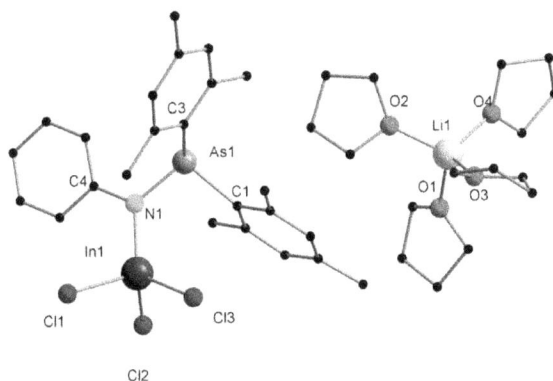

Figure 3.3.2 Molecular structure of compound **13**, [(Mes$_2$AsNPh)InCl$_3$][Li(THF)$_4$], in the solid state. Hydrogen atoms have been omitted for clarity. Selected bond lengths [Å] and angles [°]: In1-N1 2.075(2), In1-Cl3 2.3904(7), In1-Cl2 2.3885(7), In1-Cl1 2.3536(7), As1-N1 1.864(2), As1-C1 1.962(2), As1-C3 1.978(2); As1-N1-In1 126.40(11), C4-N1-As1 114.2(2), C4-N1-In1 118.8(2), Cl1-In1-Cl2 101.79(3), Cl1-In1-Cl3 107.94(3), Cl2-In1-Cl3 106.47(3), N1-As1-C1 101.05(9), N1-As1-C3 106.89(10), C1-As1-C3 108.31(11).

It should be mentioned that a slow decomposition was observed for **13** at room temperature even under inert atmosphere, resulting the formation of the aminoarsane (Mes$_2$AsN(H)Ph). Indeed, the N-H stretching absorption of Mes$_2$AsN(H)Ph was detected at 3376 cm^{-1} in the IR spectrum (Figure 3.3.3). In addition, the characteristic resonances of the methyl groups (2.35 ppm and 2.06 ppm) and NH group (3.99 ppm) of Mes$_2$AsN(H)Ph were observed in the ^{1}H NMR spectrum (Figure 3.3.4). A similar decomposition was seen in the synthesis of phosphinoamide metal complexes with the formation of the corresponding aminophosphine (R$_2$PN-(H)R').[50,69]

In summary, the mono-substituted arsinoamide complexes of aluminum (**12**) and indium (**13**) have been synthesized and structurally characterized. According to X-ray diffraction analyses, an unexpected ion pair is observed in **13**. The arsinoamide ligands in **12** and **13** adopt a *trans* conformation. However, no M-As interaction is observed in these compounds.

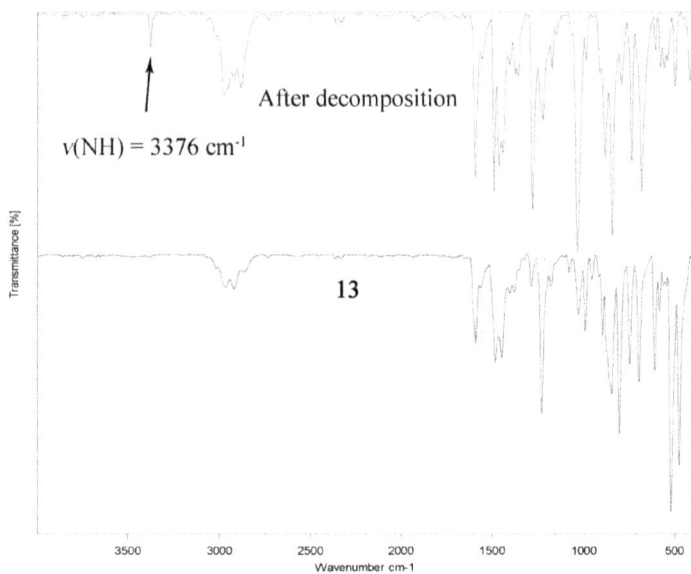

Figure 3.3.3 IR spectra of pure compound **13** and its decomposition product.

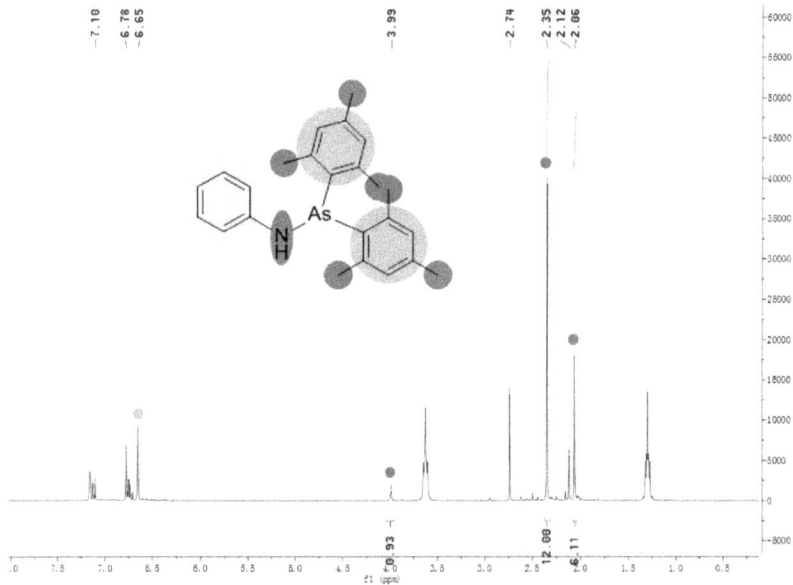

Figure 3.3.4 ^{1}H NMR spectrum of decomposed **13** in C_6D_6.

3.4 Synthesis and structural characterization of group 14 metal complexes of arsinoamide (14-21)

As mentioned before, no M-As interaction was found in the group 13 arsinoamide complexes. Thus, we moved to the low-valent group 14 elements, hoping to establish an interaction between the lone pair of As and the vacant p orbital of the low-valent group 14 elements.

The benzamidinato-supported tetrylenes [{PhC(^{t}BuN)$_2$}ECl] (E = Si and Ge) reacted at room temperature with 1 equiv. of compound **2** to give compounds **14** ([{PhC(^{t}BuN)$_2$}Si(=NPh)(AsMes$_2$)]) and **15** ([{PhC(^{t}BuN)$_2$}Ge(Mes$_2$AsNPh)]), respectively (Scheme 3.4.1). In the case of [{PhC(^{t}BuN)$_2$}SiCl], after the salt metathesis reaction, an unexpected activation of the As-N bond occurred through an oxidative addition step. Nevertheless, the heavier germylene analogue, [{PhC(^{t}BuN)$_2$}GeCl], failed to activate the As-N bond. The decreased As-N bond activation ability of the germylene can be explained by the larger singlet-triplet gap.[89,170] Detailed information about compounds **14** and **15** are given below.

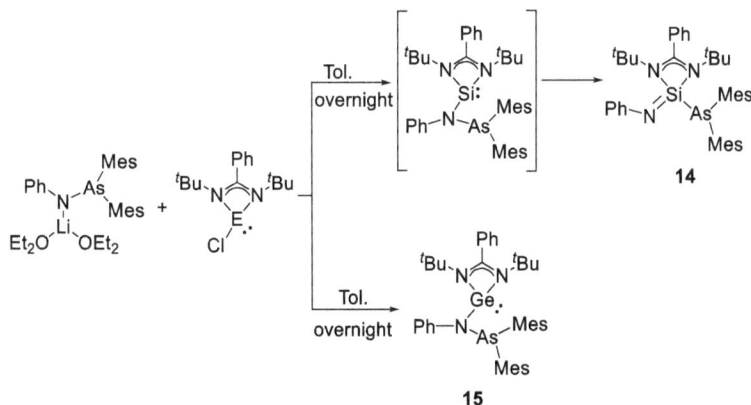

Scheme 3.4.1 Synthesis of compounds **14** ([{PhC(^{t}BuN)$_2$}Si(=NPh)(AsMes$_2$)]) and **15** ([{PhC(^{t}BuN)$_2$}Ge(Mes$_2$AsNPh)]).

Due to the insertion of the silicon atom into the As-N bond, the NMR spectra of **14** are widely altered (Figure 3.4.1 and Figure 3.4.2). In the ^{1}H NMR spectrum, the resonance of the *ortho*-methyl groups of the mesityl ring (2.83 ppm) is slightly downfield shifted compared with that of **15** at 2.65 ppm. Another remarkable difference lies in the phenyl signals of the arsinoamide ligand. Due to the different chemical environment of the nitrogen atom, narrowed and upfield shifted signals are detected for the phenyl group (7.49-6.83 ppm (**14**) *vs.* 7.79-6.86 ppm (**15**)). In the ^{13}C{^{1}H} NMR spectrum, the resonance of the *ortho*-methyl groups is shifted to 25.5 ppm (**14**) compared with 23.1 ppm for **15**. In the ^{29}Si NMR spectrum, in contrast with a signal at 14.60 ppm for [{PhC(^{t}BuN)$_2$}SiCl],[73] compound **14** gives a resonance at -68.85 ppm (Figure 3.4.3), which fits well with the published values for four-fold coordinated silicon compounds [LSi(=NAd)X] (L = PhC(^{t}BuN)$_2$, X = NPh$_2$ and NMe$_2$, Ad = adamantyl).[171]

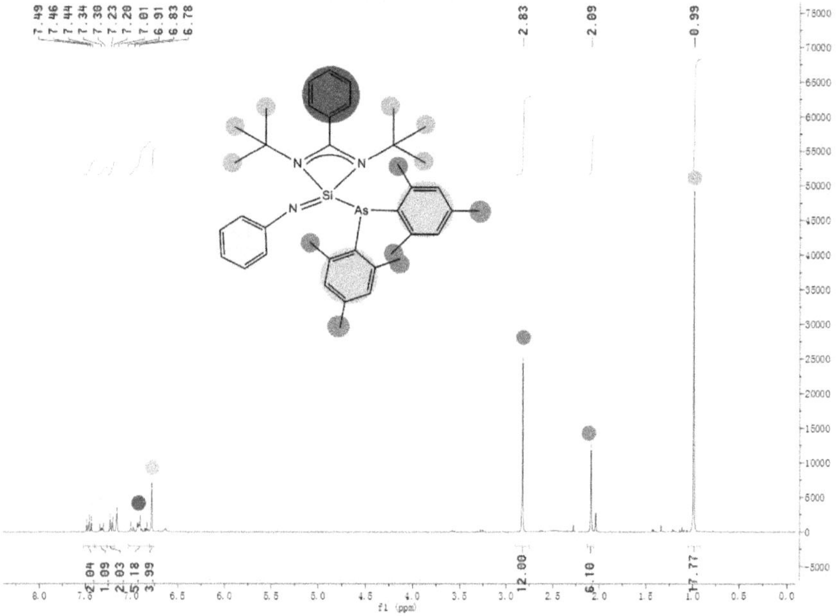

Figure 3.4.1 ^{1}H NMR spectrum of compound **14** in C$_6$D$_6$.

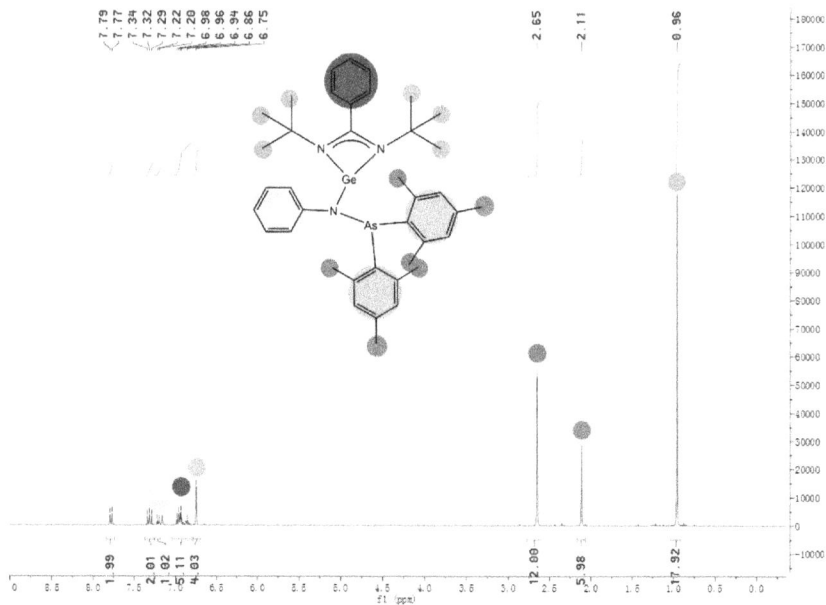

Figure 3.4.2 ^{1}H NMR spectrum of compound **15** in C$_6$D$_6$.

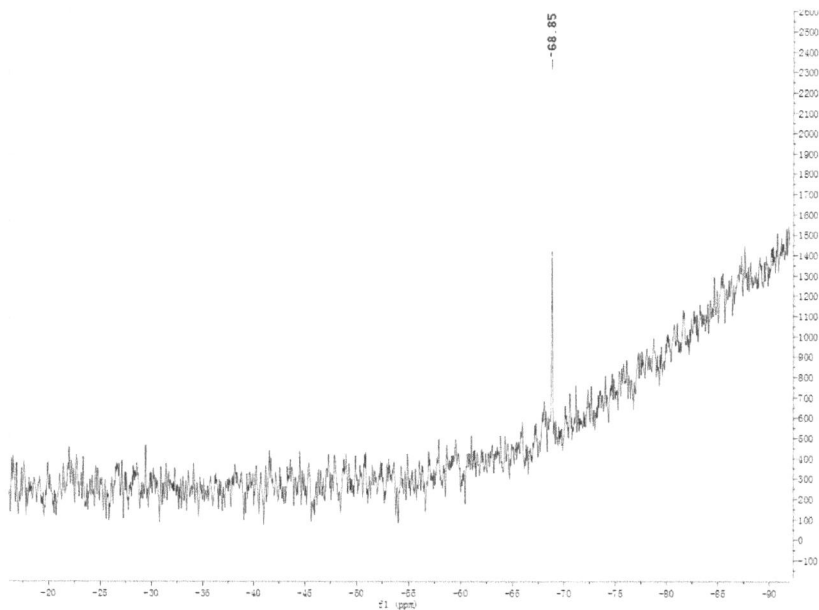

Figure 3.4.3 ^{29}Si NMR spectrum of compound **14**.

Single crystals of compound **14** were obtained from a mixture of diethyl ether and *n*-pentane (1/1) by slow evaporation. The complex crystallized in the monoclinic space group $P2_1/c$ with four molecules in the unit cell (Figure 3.4.4). As a result of the activation of the As-N bond, an unexpected Si(IV)-As(III) single bond is formed with a distance of 2.3971(7) Å, which is consistent with that of 2.352(1) Å in [Mo(CO)$_4$(Me$_2$AsSiMe$_2$CH$_2$CH$_2$AsMe$_2$)].[172] The Si1-N1 and Si1-N2 single bond lengths are 1.840(2) Å and 1.834(2) Å. However, the Si1=N3 bond length (1.571(2) Å) is in good agreement with the reported values for Si=N double bonds (*e.g.* 1.5728(12) Å or 1.584(5) Å).[171] In addition, compound **14** is the first silicon compound having an N=Si-As unit. The four-coordinated silicon atom features a distorted tetrahedral geometry, and the N3-Si1-As1 angle, formed by the insertion of Si into the As-N bond, is 133.91(9)°.

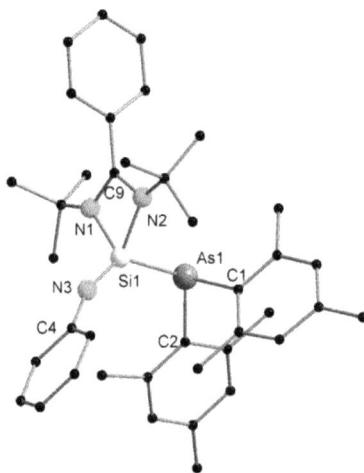

Figure 3.4.4 Molecular structure of compound **14**, [{PhC(ᵗBuN)$_2$}Si(=NPh)(AsMes$_2$)], in the solid state. Hydrogen atoms have been omitted for clarity. Selected bond lengths [Å] and angles [°]: As1-Si1 2.3971(7), As1-C1 1.978(2), As1-C2 1.986(2), Si1-N1 1.840(2), Si1-N2 1.834(2), Si1-N3 1.571(2); N1-Si1-As1 99.40(6), N2-Si1-As1 98.00(7), N2-Si1-N1 71.75(8), N3-Si1-As1 133.91(9), N3-Si1-N2 118.89(11), N3-Si1-N1 117.01(11).

While setting up this thesis, the activation of a P-N bond, through the reaction of [{PhC(ᵗBuN)$_2$}SiCl] with phosphinoamides ([DippNLiPPh$_2$]), was reported (Scheme 3.4.2).[103]

Compared with the As-N bond activation that occurred under very mild conditions (*r.t.*, overnight), the reported benzamidinato-supported silylene could activate the P-N bond only under harsher conditions (80 °C, 18 h), which may due to the double-bond character of the P-N bond.

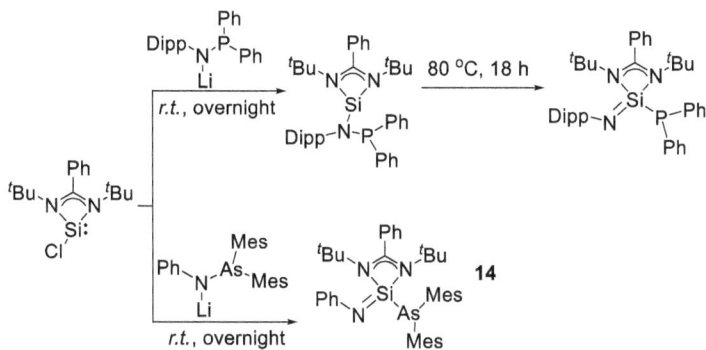

Scheme 3.4.2 Reactivity of phosphinoamides and arsinoamides with the silylene [{PhC(^{t}BuN)$_2$}SiCl].

According to the literature, some similar bond activations have already been observed. The As-H bond activation was achieved using a ylide-like silylene (Scheme 3.4.3).[173] At a temperature of -50 °C, this silylene could react with AsH$_3$ to give the 1,1-addition product, leading to the formation of a HSi-AsH$_2$ unit. Nevertheless, above -30 °C, rearrangement into an unexpected HSi=AsH unit was detected.

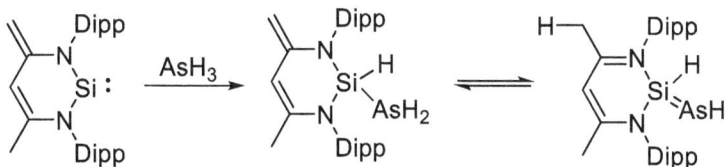

Scheme 3.4.3 Insertion of the ylide-like silylene into the As-H bond.

Only under harsh conditions, the monovalent group 13 complexes RE (R = HC{C(Me)N(2,6-^{i}Pr$_2$C$_6$H$_3$)}$_2$, E = Ga, Al) could split the As-N bond, forming the binuclear [{R(Me$_2$N)E}$_2$As$_2$] complexes (Scheme 3.4.4).[174] Compared with the reaction conditions of **14**, it seems that the nature of the metal atom and the substituents on As could significantly influence the activation energy and speed of the reaction.

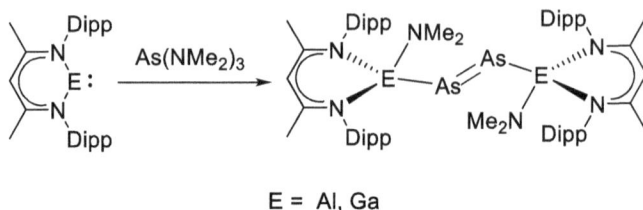

E = Al, Ga

Scheme 3.4.4 The As-N bond cleavage by the monovalent group 13 complexes.

Colorless crystals of compound **15** were obtained from *n*-pentane by slow evaporation, and the complex crystallized in the triclinic space group *P*-1 (Figure 3.4.5). Compared with the As-N separation of 1.821(7) Å in compound **2**, the As-N bond in **15** is slightly longer (1.883(2) Å). X-ray diffraction studies showed that no Ge-As interaction is present. The Ge-N bond distances range from 1.909(2) to 2.065(2) Å, which fit well with the published values for single bonds.[175-177] The sum of the bonding angles around the germanium atom (268.19°) is significantly smaller than 360°. Thus, the coordination geometry around Ge is best described as a distorted trigonal pyramidal. Due to a different nature of the metal atom, the Ge1-N1-As1 angle in **15** (120.65(10)°) is increased compared with the Li-N-As angle in compound **2** (113.6(5)°).[135]

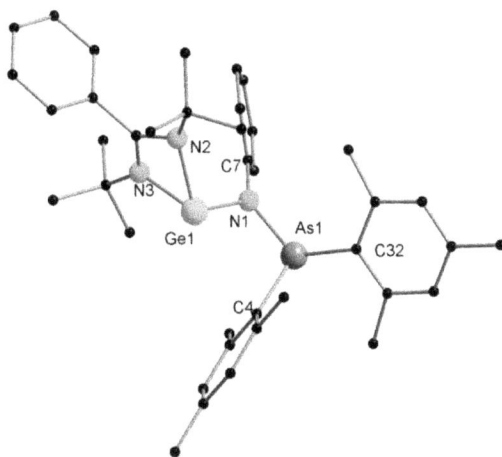

Figure 3.4.5 Molecular structure of compound **15**, [{PhC(^{t}BuN)$_2$}Ge(Mes$_2$AsNPh)], in the solid state. Hydrogen atoms have been omitted for clarity. Selected bond lengths [Å] and angles [°]: As1-N1 1.883(2), As1-C4 1.979(2), As1-C32 1.987(2), Ge1-N1 1.909(2), Ge1-N2 2.065(2), Ge1-N3 2.001(2), N1-C7 1.430(3); N1-As1-C4 99.04(9), N1-As1-C32 109.21(9), C4-As1-C32 103.57(10), N1-Ge1-N2 105.29(8), N1-Ge1-N3 98.67(8), N3-Ge1-N2 64.23(8), As1-N1-Ge1 120.65(10), C7-N1-As1 113.1(2), C7-N1-Ge1 122.8 (2), N3-C10-N2 108.8(2).

To further understand the As-N bond activation by tetrylenes, GeCl$_2$·dioxane, without any supporting ligand, was utilized to react with compound **2** in 1:1 ratio (Scheme 3.4.5). According to X-ray diffraction studies, an As-N bond activation was detected again in compound **16**. The germanium atom inserts into the As-N bond, after oxidative addition reaction, forming a Ge$_2$N$_2$ four-membered ring. Compared with [{PhC(^{t}BuN)$_2$}GeCl], the observed reactivity of the intermediate germylene may be affected by the nature of the substituents on the germanium atom.[178-180] Motivated by the reactivity of GeCl$_2$·dioxane with arsinoamide, the reaction of SnCl$_2$ and PbCl$_2$ with 1 equiv. of compound **2** was investigated, resulting in the formation of compounds **17** ([(Mes$_2$AsNPh)SnCl{THF}]) and **20** ([(Mes$_2$AsNPh)$_2$Pb]) (Scheme 3.4.5). However, X-ray diffraction studies revealed no As-N bond activation. Due to the inert pair effect, the ability to activate σ-bonds usually decreases from Si to Pb.[91] Still, a remarkable difference lies between the synthesis of compounds **17** and **20**. Using SnCl$_2$, the salt metathesis reaction resulted in the formation of the mono-substituted compound **17**. Due to the presence of both a chloride ligand and a lone pair on the arsenic atom, compound **17** may exhibit a dual reactivity, leading to either the substitution of the chloride or the coordination of the arsenic atom. Nevertheless, due to the larger ionic radius of the lead atom, the bis-substituted compound **20** was isolated, even when using 1 equiv. of **2**. It seems that only one arsinoamide ligand is not able to provide enough steric protection and/or electronic stabilization of the lead(II) center.[181-183] Detailed information about compounds **16**, **17** and **20** are given below.

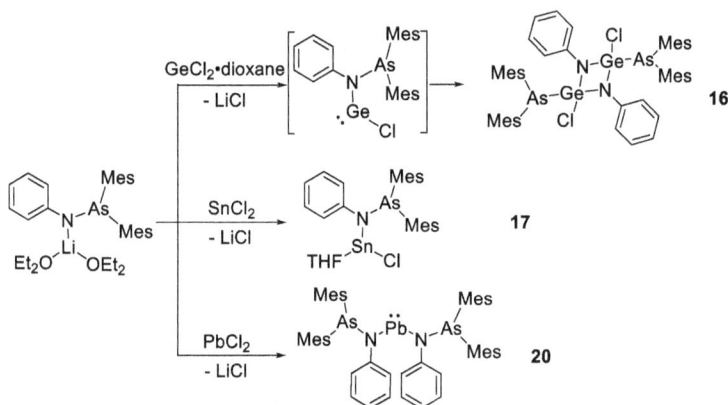

Scheme 3.4.5 Synthesis of compounds **16**, **17** and **20**.

Due to the insertion of the germanium atom into the As-N bond, the resonance of the *para*-methyl groups of **16** is shifted to 2.22 ppm, compared with 2.05 ppm in **17** in the ^{1}H NMR spectrum (Figure 3.4.6). In addition, an up-field shifted of the phenyl signals could be seen (6.81-6.49 ppm (**16**) *vs.* 7.13-6.80 ppm (**17**)). In the case of compound **20**, the resonances of the *ortho*- and *para*-methyl groups are detected at 2.34 and 2.04 ppm, respectively. In the ^{119}Sn NMR spectrum, the resonance of **17** is observed at -34.2 ppm (Figure 3.4.7), which is in the similar range as that of [{(SiMe$_3$)$_2$N}C(^{i}PrN)$_2$SnCl] at -51.1 ppm and of [{2-[(CH$_3$)$_2$NCH$_2$]C$_6$H$_4$}(SiMe$_3$)NSnCl] at -51.2 ppm.[184,185]

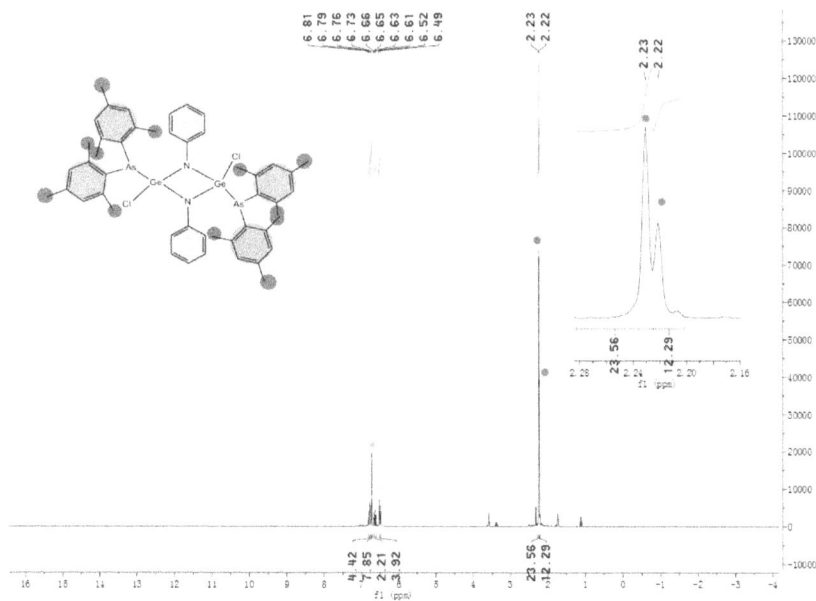

6.81
6.79
6.76
6.73
6.66
6.65
6.63
6.61
6.52
6.49

2.23
2.22

2.23
2.22

4.42
7.85
2.21
3.92

23.56
12.29

23.56
12.29

Figure 3.4.6 ^{1}H NMR spectrum of compound **16** in d_8-THF.

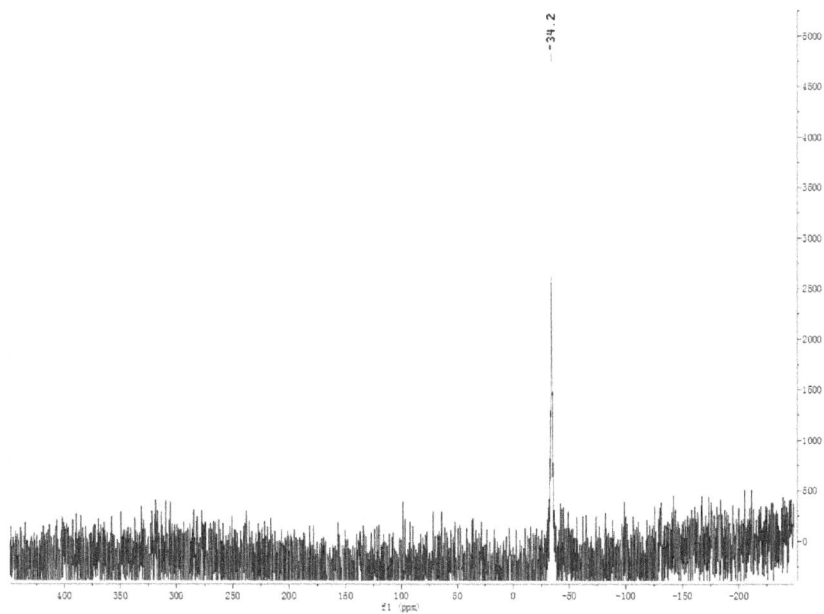

-34.2

Figure 3.4.7 ^{119}Sn NMR spectrum of compound **17**.

53

Single crystals of compound **16** were isolated from a mixture of diethyl ether and *n*-pentane (1/1) stored at -30 °C, and the complex crystallized in the monoclinic space group $P2_1/c$ (Figure 3.4.8). X-ray diffraction analyses indicate that **16** reveals a dimeric configuration with a planar Ge_2N_2 core [186-188]. The Mes_2As groups arranged in a *trans* fashion across the ring. A rare Ge(IV)-As(III) bond is detected with a distance of 2.4327(9) Å, which agrees with that of 2.457(1) Å in 1,3-diarsa-4-sila-2-germacyclobutanes and of 2.444(2) Å in lithioarsinoorganogermanes. [189,190] The Ge-N bond length (1.847(4) Å) is in the reported range for Ge-N single bonds. [191-193] The four-fold coordinated germanium center exhibits a distorted tetrahedral geometry.

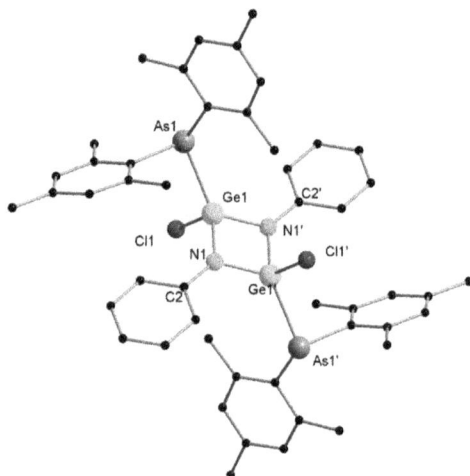

Figure 3.4.8 Molecular structure of compound **16**, [{Mes_2As}ClGe(μ-NPh)]$_2$, in the solid state. Hydrogen atoms have been omitted for clarity. Selected bond lengths [Å] and angles [°]: Ge1-As1 2.4327(9), Ge1-Cl1 2.1993(14), Ge1-N1 1.847(4), N1-C2 1.392(6); N1-Ge1-As1 126.30(14), N1-Ge1-N1' 83.8(2), N1-Ge1-Cl1 107.97(14), Ge1-N1-Ge1' 96.2(2).

The As-N bond activation by $GeCl_2$·dioxane is unexpected. Besides, no P-N bond activation by germylene has ever been reported (Chapter 1.2.4). It should be mentioned that in the described synthesis of silylene-germylene phosphinoamide complexes, decomposition of the phosphinoamide ligand was detected, resulting in the formation of a mixed-valent complex in very low yield (4.9 %) (Scheme 3.4.6) [103].

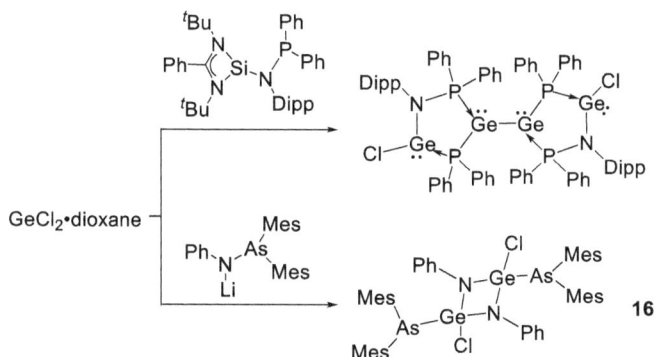

Scheme 3.4.6 Reactivity of GeCl$_2$·dioxane towards phosphinoamide and arsinoamide.

Single crystals of **17** were grown from a mixture solution of tetrahydrofuran and diethyl ether (1/1) at -30 °C, and the complex crystallized in the triclinic space group *P*-1 with two molecules in the unit cell (Figure 3.4.9). The Sn-N (2.098(3) Å) and Sn-Cl bond lengths (2.4587(12) Å) are consistent with the published values (*e.g.* 2.114(3) Å (Sn-N) or 2.444(1) Å (Sn-Cl)).[194] Due to the different nature of the metal atom, the As-N bond length in **17** is a little longer than that in compound **2** (1.875(3) Å in **17** *vs* 1.821(7) Å in **2**). However, the As-N-M angles are similar (As1-N1-Sn1: 114.67(16)° *vs* As1-N1-Li1: 113.6(5)°). Considering the sum of the angles around the tin atom (280.42°), the three-fold coordinated Sn exhibits a distorted trigonal pyramidal geometry. According to X-ray diffraction analyses, no Sn-As interaction is observed in compound **17**.

Detailed information about compound **20** is given below.

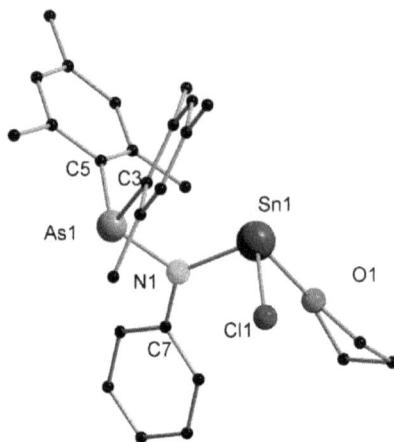

Figure 3.4.9 Molecular structure of compound **17**, [(Mes₂AsNPh)SnCl(THF)], in the solid state. Hydrogen atoms have been omitted for clarity. Selected bond lengths [Å] and angles [°]: Sn1-Cl1 2.4587(12), Sn1-O1 2.307(3), Sn1-N1 2.098(3), As1-N1 1.875(3), As1-C3 1.991(4), As1-C5 1.973(4); O1-Sn1-Cl1 88.92(8), N1-Sn1-Cl1 99.02(10), N1-Sn1-O1 92.48(12), N1-As1-C3 95.9(2), N1-As1-C5 110.9(2), As1-N1-Sn1 114.7(2), C7-N1-Sn1 130.5(2), C7-N1-As1 113.9(2).

Encouraged by the rich reactivity of the mono-substituted complexes, we moved to the synthesis of the homoleptic bis-substituted complexes, [(Mes₂AsNPh)₂E] (E = Ge (**18**), Sn (**19**), Pb (**20**)), by reaction of ECl_2 with a stoichiometric amount of compound **2** (Scheme 3.4.7). One interesting feature is the color of these compounds in the solid state: bright yellow (**18**), red (**19**) or deep violet (**20**). In addition, it is worth noting that, in a similar manner to the dialkyl stannylene ([{(SiMe₃)₂CH}₂Sn]) [28], the color of compound **19** could reversibly turn to orange at -30 °C and further to yellow at liquid-nitrogen temperature (-196 °C). This striking diversification may be related to an intramolecular charge transfer mechanism.[195,196]

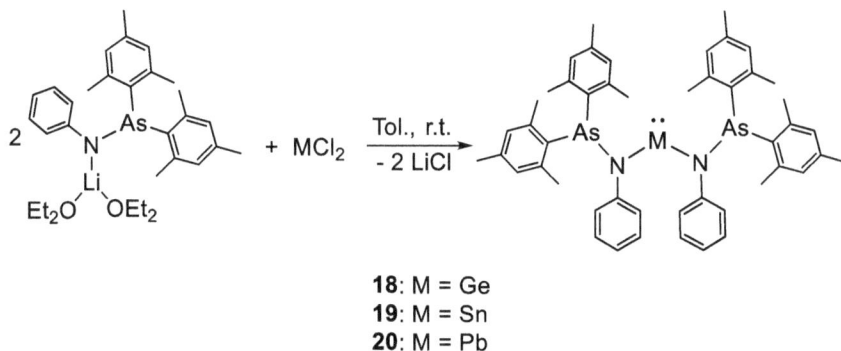

18: M = Ge
19: M = Sn
20: M = Pb

Scheme 3.4.7 Synthesis of compounds **18-20**.

X-ray diffraction analyses reveal that compounds **18**, **19** and **20** are isostructural and none of these compounds features a M-As interaction. As expected, the E-N bond lengths increase from 1.8791(15) Å (**18**) to 2.125(4) Å (**19**) and 2.222(6) Å (**20**) because of the increasing ionic radii. Moreover, the N-E-N bond angles decrease from 101.33(10)° (**18**) to 97.93(14)° (**19**) and 97.4(2)° (**20**). A consistent trend can also be found in the dimeric amido tetrylene complexes.[186] Selected bond lengths and angles of compounds **18-20** are listed in Table 3.4.1. Detailed structural information about compounds **18-20** are given below.

Table 3.4.1 Selected bond lengths and angles of compounds **18-20**.

Compound	M-N bond length (Å) [a]	N-M-N angle (°)
[(Mes$_2$AsNPh)$_2$Ge] (**18**)	1.879(2)	101.33(10)
[(Mes$_2$AsNPh)$_2$Sn] (**19**)	2.128(4)	97.93(14)
[(Mes$_2$AsNPh)$_2$Pb] (**20**)	2.222(6)	97.4(2)

[a]: average bond length

In the solid state, compounds **18**, **19** and **20** are stable at room temperature under inert atmosphere. Nevertheless, a slow decomposition could be detected in C$_6$D$_6$ at room temperature. Compounds **18**, **19** and **20** manifest similar ^{1}H and ^{13}C{^{1}H} NMR spectra. In the ^{119}Sn NMR spectrum, the resonance of **19** is detected at 318.6 ppm (Figure 3.4.10), which agrees with that at 267.9 ppm for the bis-substituted stannylene.[197] In the case of **20**, the divalent diamido plumbylene signal is detected at 3244 ppm in the ^{207}Pb NMR spectrum (Figure 3.4.11), which may be compared with a resonance at 3504 ppm for the diamido plumbylene compound [{DippN(CH$_2$)$_3$NDipp}Pb].[198]

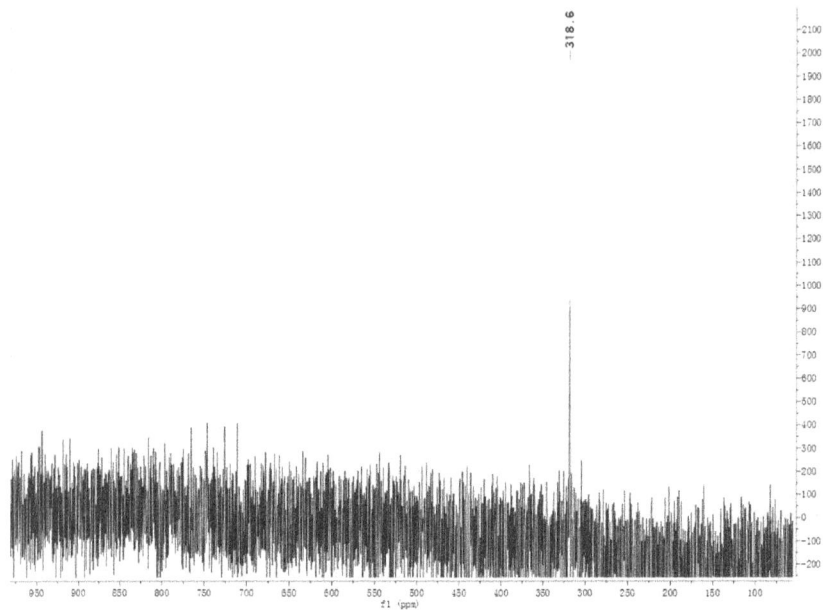

Figure 3.4.10 ^{119}Sn NMR spectrum of compound **19**.

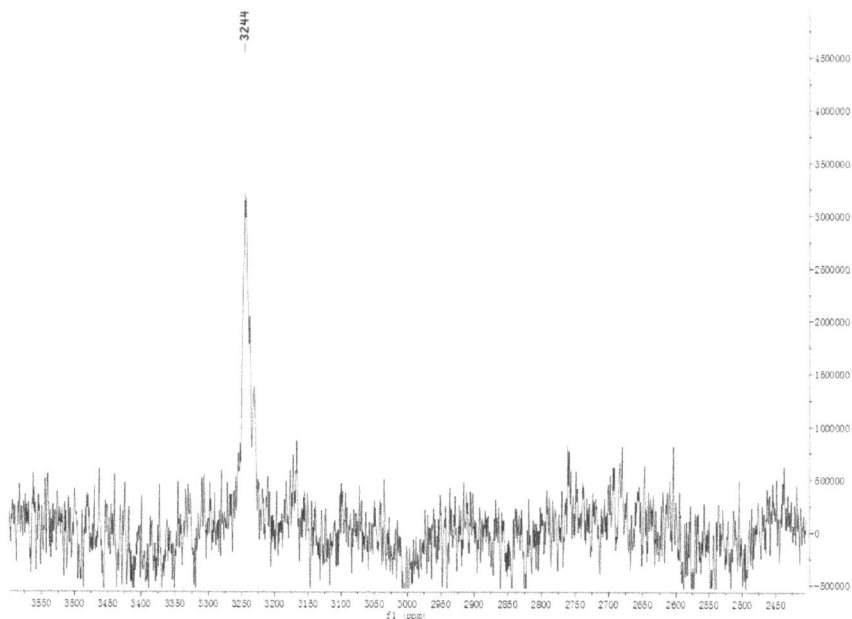

Figure 3.4.11 ^{207}Pb NMR spectrum of compound **20**.

After the reaction of GeCl$_2$·dioxane with 2 equiv. of compound **2** in toluene, compound **18** was isolated as a yellow solid with a good yield (87 %). Single crystals of **18** were obtained from an *n*-heptane solution stored at -30 °C, and the complex crystallized in the monoclinic space group *C2/c* with four molecules in the unit cell (Figure 3.4.12). The Ge-N bond distance is 1.879(2) Å, which fits well with the reported values.[199,200] Considering the sum of the bonding angles (359.45°), the nitrogen atoms adopt a slightly distorted trigonal planar coordination geometry. The arsinoamides exhibit a *trans* conformation, while a *cis* conformation was observed in germylene phosphinoamide complexes.[105] Due to the different conformation of the ligands, the N1-Ge-N1' angle (101.33(10)°) in **18** is smaller than that observed in the bis(phosphinoamide)germylene complex (107.1°).[105] This difference may relate to the steric effect of the ligands.[25] According to the X-ray diffraction studies, no Ge-As interaction was observed in compound **18**. Similar as the phosphinoamides zirconium complex,[37,45,69,70] compound **18** provide an alternative route to synthesis the heterobimetallic complexes, thanks to the presence of lone pair on both As and Ge.

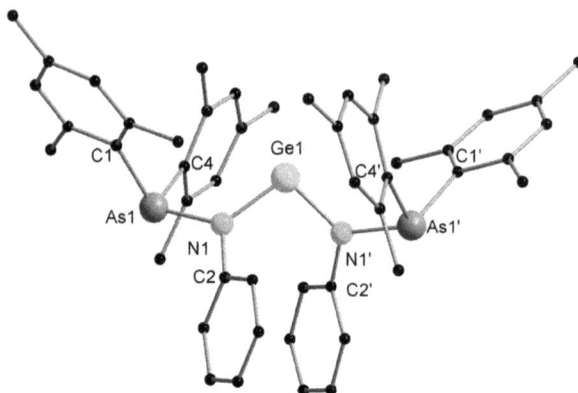

Figure 3.4.12 Molecular structure of compound **18**, [(Mes₂AsNPh)₂Ge], in the solid state. Hydrogen atoms have been omitted for clarity. Selected bond lengths [Å] and angles [°]: As1-N1 1.879(2), As1-C1 1.982(2), As1-C4 1.990(2), Ge1-N1 1.879(2), N1-C2 1.425(2); N1-As1-C1 109.47(7), N1-Ge1-N1' 101.33(10), As1-N1-Ge1 121.92(8), C2-N1-Ge1 122.03(11), C2-N1-As1 115.50(11).

Transmetallation of $SnCl_2$ with 2 equiv. of compound **2** in toluene led to compound **19** [(Mes₂AsNPh)₂Sn] as a red solid. Single crystals of compound **19** were grown by diffusion of a mixture of *n*-pentane and diethyl ether (1/1) into a solution of **19** in toluene. The complex crystallized in the tetragonal space group $I4_1/a$ (Figure 3.4.13). The arsinoamides exhibit a *trans* conformation. The As-N bond lengths are similar to that in compound **17** (1.877(4) and 1.870(4) Å (**19**) *vs* 1.875(3) Å (**17**)). The Sn-N bond distances (2.125(4) and 2.131(4) Å) fit well with the published values for Sn-N single bonds.[184,201,202] Considering the sum of the bonding angles (356.3° and 350.8°), the three-fold coordinated nitrogen atoms feature a distorted trigonal planar coordination environment. The N-M-N bond angle (N1-Sn1-N2: 97.93(14)°) is less than that of 101.33(10)° (N1-Ge1-N1') in compound **18**. According to X-ray diffraction studies, no Sn-As interaction is observed in compound **19**.

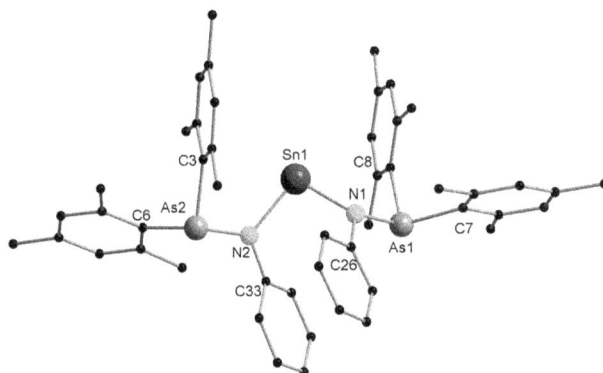

Figure 3.4.13 Molecular structure of compound **19**, [(Mes₂AsNPh)₂Sn], in the solid state.
Hydrogen atoms and *n*-pentane molecular have been omitted for clarity. Selected bond lengths
[Å] and angles [°]: Sn1-N1 2.125(4), Sn1-N2 2.131(4), As1-N1 1.877(4), As1-C7 2.000(4),
As1-C8 1.980(5), As2-N2 1.870(4), As2-C3 1.981(4), As2-C6 1.995(4), N1-C26 1.412(5),
N2-C33 1.412(6); N2-Sn1-N1 97.93(14), N1-As1-C7 109.8(2), N1-As1-C8 97.1(2), C8-As1-C7
99.8(2), N2-As2-C3 97.3(2), As1-N1-Sn1 115.8(2), As2-N2-Sn1 121.4(2), C33-N2-Sn1
118.0(3).

After the reaction of PbCl₂ with 2 equiv. of compound **2** in toluene, compound **20** was isolated as
a deep violet solid. It should be noted that compound **20** is extremely sensitive towards moisture
and air. Single crystals of compound **20** were grown by slowly diffusing a mixture of *n*-pentane
and diethyl ether (1/1) into a solution of **20** in toluene. The complex crystallized in the tetragonal
space group *I*4₁/*a* (Figure 3.4.14). According to the X-ray diffraction studies, compounds **20** and
19 are isostructural. The Pb-N bond lengths, established by X-ray diffraction studies, are of
2.236(6) Å and 2.208(6) Å, which are in line with the corresponding published single bond
values.[181,198,203] The nitrogen atoms are three-fold coordinated and in a distorted trigonal planar
coordination geometry (sums of the angles around the nitrogen atoms of 354.5° and 350.7°). In
addition, compared with the germylene and stannylene analogues, the N1-Pb-N2 angle (97.4(2)°)
is lower than that of 101.33(10)° and 97.93(14)° in compounds **18** (N-Ge-N) and **19** (N-Sn-N),
respectively. The arsinoamides exhibit a *trans* conformation. According to the crystal structure,
no Pb-As interaction is observed in compound **20**. To the best of our knowledge, no comparable
structurally characterized plumbylene phosphinoamide complex has ever been reported.

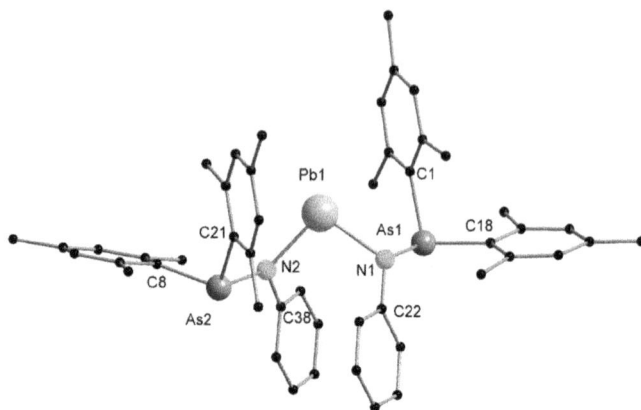

Figure 3.4.14 Molecular structure of compound **20**, [(Mes₂AsNPh)₂Pb], in the solid state. Hydrogen atoms and *n*-pentane molecular have been omitted for clarity. Selected bond lengths [Å] and angles [°]: Pb1-N1 2.236(6), Pb1-N2 2.208(6), As1-N1 1.864(6), As1-C1 1.990(7), As1-C18 1.996(7), As2-N2 1.873(7), As2-C8 1.986(8), As2-C21 1.965(8), N1-C22 1.401(9), N2-C38 1.418(10); N2-Pb1-N1 97.4(2), N1-As1-C1 97.1(3), N1-As1-C18 108.8(3), C1-As1-C18 100.5(3), N2-As2-C8 108.7(3), As1-N1-Pb1 114.1(3), As2-N2-Pb1 121.5(3), C38-N2-Pb1 116.8(5), C38-N2-As2 116.2(5).

In summary, no As-N bond activation can be observed in compounds **18** and **19**, which is consistent with the reported successful synthesis of phosphinoamide germylene and stannylene complexes (Chapter 1.2.4).[101,104-106]

As potential bridging ligands, phosphinoamides have been widely used to synthesize bimetallic complexes.[9,50,204,205] The heavier arsinoamide analogues should also be able to act as bridging ligands. Thus, treatment of compound **18** with [Ni(COD)₂] (COD: 1,5-cyclooctadiene) resulted in the first arsinoamide-supported Ge/Ni complex **21** [(Mes₂AsNPh)₂GeNi(COD)] (Scheme 3.4.8). Due to the decomposition of the complex in deuterated solvents, no clean NMR spectra of **21** can be recorded.

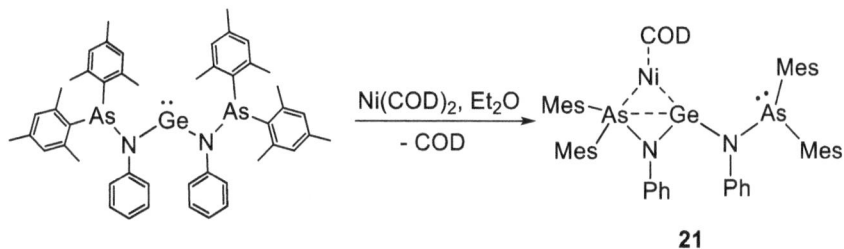

Scheme 3.4.8 Synthesis of arsinoamide-supported Ge/Ni complex (**21**).

Single crystals of compound **21** were grown from a diethyl ether solution stored at -30 °C, and the complex crystallized in the triclinic space group P-1 (Figure 3.4.15). After the leaving of one COD molecule, the nickel center coordinates to one germanium and one arsenic atom with bond distances of 2.227(2) Å (Ge-Ni) and 2.352(2) Å (Ni-As).[206,207] Similar bond length values could be found in the published literature, $e.g.$ 2.263 Å for the Ni-As bond length in [(Ph$_3$As)Ni(μ-PPh$_2$)$_3$Ni(AsPh$_3$)][208] or 2.240 Å for the Ge-Ni bond length in [Ni{Ge[(^{i}PrN)$_2$C$_{10}$H$_6$]}$_4$].[209] In addition, the As2-N2-Ge1 angles decreases sharply from 120.4(4)° (**18**) to 95.2(4)° (**21**). As a result, a weak Ge1···As2 interaction (2.7910(14) Å) can be detected, taking into account the longest reported Ge-As interaction of 2.816(12) Å in [{Ta@Ge$_8$As$_6$}$^{3-}$].[210] The N1-Ge1-N2 bond angle is similar to that in **18** (angle of 105.7(3)° in **21** and 101.33(10)° in **18**). In addition, the conformational interconversion of one arsinoamide ligand (from $trans$ in **18** to cis in **21**) was detected in **21**.

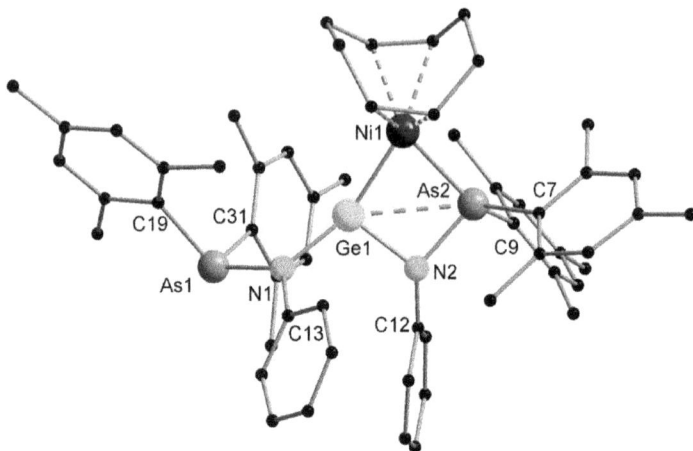

Figure 3.4.15 Molecular structure of compound **21**, [(Mes$_2$AsNPh)$_2$GeNi(COD)], in the solid state. Hydrogen atoms have been omitted for clarity. Selected bond lengths [Å] and angles [°]: As1-N1 1.855(8), As1-C19 1.977(10), As1-C31 1.960(9), Ge1-As2 2.7910(14), Ge1-Ni1 2.227(2), Ge1-N1 1.884(7), Ge1-N2 1.907(8), As2-Ni1 2.352(2), As2-N2 1.874(8), As2-C7 1.960(9), As2-C9 1.960(10); N1-As1-C19 108.8(4), N1-As1-C31 96.6(4), C31-As1-C19 105.6(4), Ni1-Ge1-As2 54.50(4), N1-Ge1-As2 146.8(3), N1-Ge1-Ni1 150.0(2), N1-Ge1-N2 105.7(3), N2-Ge1-As2 42.0(2), N2-Ge1-Ni1 95.4(3), Ni1-As2-Ge1 50.44(4), N2-As2-Ge1 42.9(3), N2-As2-Ni1 92.3(3), N2-As2-C7 99.6(4), N2-As2-C9 114.9(4), Ge1-Ni1-As2 75.07(5), As1-N1-Ge1 120.4(4), As2-N2-Ge1 95.2(4).

4 Experimental section

All manipulations of air-sensitive materials were performed under rigorous exclusion of oxygen and moisture in Schlenk-type glassware or on a dual manifold Schlenk line, interfaced to a high vacuum (10^{-3} torr) line, or in an argon-filled MBraun glove box. Hydrocarbon solvents (toluene, n-pentane, n-heptane) and diethyl ether were predried using an MBraun solvent purification system (SPS-800) and then they were degassed, dried and stored in vacuo over $LiAlH_4$. Tetrahydrofuran was distilled under nitrogen from potassium before storage over $LiAlH_4$. Dichloromethane was distilled under nitrogen from CaH_2 prior to use. Deuterated solvents (C_6D_6, d_8-THF) were obtained from Euriso-Top (99.5 atom % D) and were degassed, dried and stored in vacuo over Na/K alloy in resealable flasks. IR spectra were obtained on a Bruker Tensor 37 instrument equipped with a room temperature DLaTGS detector, a diamond ATR unit and a nitrogen flushed chamber. The ^1H, ^{13}C, ^{29}Si, ^{119}Sn, and ^{207}Pb NMR spectra were recorded on a Bruker Avance II 300 MHz or Avance III 400 MHz NMR spectrometer. Chemical shifts are expressed in parts per million (ppm) and referenced to the residual ^1H and ^{13}C resonances of the deuterated solvents and are reported relative to tetramethylsilane. The ^{119}Sn and ^{207}Pb NMR data were referenced to $SnMe_4$ and $PbMe_4$, respectively. Mass spectra were recorded at 70 eV on a Thermo Fisher Scientific DFS – Magnetic SectorGC/MS HRMS instrument. Elemental analyses were carried out with an Elementar Vario micro cube (Elementar Analysensysteme GmbH).

Triethylamine was dried by CaH_2 and distilled before use [211]. Aniline was freshly distilled prior to use. Mes_2AsCl,[123] $[(NMe_2)_2ZrCl_2 \cdot (THF)_2]$,[149] $[(NMe_2)_2HfCl_2 \cdot (THF)_2]$,[149] $[\{PhC(N^tBu)_2\}SiCl]$[73] and $[\{PhC(N^tBu)_2\}GeCl]$[75] were prepared according to the literature procedures. The other chemicals were purchased from Abcr, Acros Organics, Alfa Aesar, Sigma Aldrich, TCI and used without further purification.

4.1 Synthesis of aminoarsane and alkali metal complexes of arsinoamide

4.1.1 Mes$_2$AsN(H)Ph (1)

To a stirred solution of Mes$_2$AsCl (6.98 g, 20.0 mmol) and aniline (1.8 mL, 20.0 mmol) in THF (150 mL), triethylamine (2.8 mL, 20.0 mmol) was added dropwise. After stirring overnight at room temperature, the white precipitate that formed was removed by filtration. The product was isolated as a white powder after removing of the volatiles under vacuum and washing with cold n-pentane. Yield: 6.57 g, (81 %).

^{1}H NMR (400 MHz, C$_6$D$_6$, 298 K): δ(ppm) = 7.15-7.11 (m, 2H, m-H Ph), 6.77-6.72 (m, 3H, o,p-H Ph), 6.65 (s, 4H, ring CH Mes), 3.98(s, 1H, NH), 2.35(s, 12H, o-CH$_3$ Mes), 2.06(s, 6H, p-CH$_3$ Mes).

^{13}C{^{1}H} NMR (101 MHz, C$_6$D$_6$, 298 K): δ(ppm) = 149.8(Ph), 141.6(Mes), 138.4(Mes), 138.3(Mes), 130.4(Mes), 129.6(Mes), 129.6(Ph), 118.2(Ph), 116.1(Ph), 22.5 (o-CH$_3$ Mes), 20.9 (p-CH$_3$ Mes).

IR (ATR, cm^{-1}): 3377(m), 3023(w), 2969(w), 2922(w), 2859(w), 1595(s), 1493(s), 1465(s), 1442(s), 1361(m), 1283(s), 1227(m), 1175(m), 1029(m), 848(s), 743(s), 689(s), 613(m), 584(m), 563(m), 502(m), 419(w).

MS (EI, 70 eV): m/z (%) 405 ([M]$^+$, 51), 312 ([M-(PhNH)]$^+$, 100), 91 ([M-(Mes$_2$As)]$^+$, 37).

Anal. calcd. (%) for [C$_{24}$H$_{28}$AsN] (405.42): C, 71.10; H, 6.96; N, 3.45. Found: C, 71.17; H, 7.19; N, 2.68.

4.1.2 [(Mes$_2$AsNPh){Li(OEt$_2$)$_2$}] (2)

At a temperature below -78 °C, n-BuLi (2.5 M in n-hexane, 2.0 mL, 5.0 mmol) was added dropwise to a stirred solution of **1** (2.03 g, 5.0 mmol) in THF (50 mL). After warming to room temperature, the mixture turned light yellow. It was kept stirring for further 6 h. The resulting solution was filtered, and the volatiles were removed in vacuum. The residue was washed with n-pentane (3 x 20 mL) to give a white powder. Single crystals suitable for X-ray diffraction were obtained from a THF/diethyl ether (1/1) solution at -30 °C after 5 days. Yield: 2.1 g (75%).

^{1}H NMR (300 MHz, C$_6$D$_6$, 298 K): δ(ppm) = 7.33-7.28 (m, 2H, *m*-H Ph), 6.88-6.86 (d, 2H, *o*-H Ph), 6.72 (s, 4H, ring CH Mes), 6.68-6.63 (t, 1H, *p*-H Ph), 2.66(s, 12H, *o*-CH$_3$ Mes), 2.12(s, 6H, *p*-CH$_3$ Mes).

^{13}C{^{1}H} NMR (75 MHz, C$_6$D$_6$, 298 K): δ(ppm) = 142.7 (Ph), 136.7 (Mes), 130.4(Mes), 129.9 (Mes), 129.6(Mes), 129.5(Mes), 118.5(Ph), 116.1(Ph), 22.0 (*o*-CH$_3$ Mes), 21.0 (*p*-CH$_3$ Mes).

IR (ATR, cm^{-1}): 3016(w), 2954(w), 2919(w), 2809(w), 1589(s), 1497(s), 1480(s), 1377(w), 1261(s), 1194(m), 1065(m),1072(m), 849(m), 749(s), 692(s), 584(w), 549(m), 500(m).

Anal. calcd. (%) for [C$_{24}$H$_{27}$AsLiN·Et$_2$O, = **2** - Et$_2$O] (485.47): C, 69.27; H, 7.68; N, 2.89. Found: C, 70.03; H, 7.44; N, 2.99.

4.1.3 [(Mes$_2$AsNPh){Na(OEt$_2$)}]$_2$ (**3**)

THF (25 mL) was condensed at -78 °C to a mixture of **1** (2.03 g, 5.0 mmol) and NaN(SiMe$_3$)$_2$ (0.92 g, 5.0 mmol). After warming to room temperature, the mixture was stirred at room temperature overnight. The suspension was filtered and the volatiles were removed under vacuum. The residue was washed with *n*-pentane (3x20 mL) and a pale yellow powder was obtained. Single crystals suitable for X-ray diffraction were grown by layering *n*-heptane onto a solution of **3** in diethyl ether. Yield: 1.9 g (76%).

^{1}H NMR (300 MHz, *d$_8$*-THF, 298 K): δ(ppm) = 6.66-6.61 (m, 2H, *m*-H in Ph), 6.57 (s, 4H, ring CH Mes), 6.31-6.29 (d, 2H, *o*-H Ph), 5.86 (s, 1H, *p*-H Ph), 2.37 (s, 12H, *o*-CH$_3$ Mes), 2.11 (s, 6H, *p*-CH$_3$ Mes).

^{13}C{^{1}H} NMR (75 MHz, *d$_8$*-THF, 298 K): δ(ppm) = 166.4 (Ph), 144.7 (Mes), 143.4 (Mes), 136.4 (Ph), 129.8 (Mes), 128.8 (Mes), 118.9 (Ph), 108.6 (Ph), 21.9 (*o*-CH$_3$ Mes), 21.2 (*p*-CH$_3$ Mes).

IR (ATR, cm^{-1}): 2977 (w), 2951 (w), 2917 (w), 2868 (w), 1572 (s), 1467 (s), 1369 (m), 1304 (s), 1295 (s), 1256 (m), 1177 (m), 1049 (m), 1019 (w), 978 (s), 856 (s), 842 (s), 748 (s), 692 (s), 689 (s), 583 (m), 553 (m), 520 (m), 503 (m). **Anal. calcd.** (%) for [$C_{24}H_{27}AsNNa \cdot Et_2O$] (501.52): C, 67.06; H, 7.44; N, 2.79. Found: C, 67.56; H, 7.35; N, 2.53.

4.1.4 [(Mes$_2$AsNPh){K(THF)}]$_2$ (4)

Method 1: THF (25 mL) was condensed at -78 °C to a mixture of **1** (2.03 g, 5.0 mmol) and KH (0.20 g, 5.0 mmol). The reaction mixture was allowed to warm to room temperature and stirred at room temperature overnight, and the mixture turned brown. The mixture was filtered and the volatiles were removed under vacuum. The residue was washed with *n*-pentane (3x20mL) to give **4** as a yellow powder. Single crystals suitable for X-ray diffraction were obtained by layering *n*-pentane onto a solution of **4** in THF. Yield: 2.1 g (82%).

Method 2: THF (25 mL) was condensed at -78 °C to a mixture of **1** (2.03 g, 5.0 mmol) and KBTSA (1.00 g, 5.0 mmol). After warming to room temperature, the mixture was stirred overnight, the mixture turned brown. The mixture was filtered and the volatiles were removed in vacuum. The residue was washed with *n*-pentane (3x20mL) and diethyl ether (15 mL) to give **4** as a yellow powder. Single crystals of compound **4** could also be obtained by slow evaporation of a toluene solution. Yield: 2.0 g (81%).

^{1}**H NMR** (300 MHz, d_8-THF, 298 K): δ(ppm) = 6.67-6.62(m, 2H, *m*-H Ph), 6.59(s, 4H, ring CH Mes), 6.36-6.33 (d, 2H, *o*-H Ph), 5.83(s, 1H, *p*-H Ph), 2.39(s, 12H, *o*-CH$_3$ Mes), 2.12(s, 6H, *p*-CH$_3$ Mes).

^{13}**C{^{1}H} NMR** (75 MHz, d_8-THF, 298 K): δ(ppm) = 143.2(Ph), 142.2(Mes), 136.7 (Mes), 129.8 (Mes), 129.5 (Mes), 129.2(Mes), 118.3 (Ph), 115.1 (Ph), 22.2 (*o*-CH$_3$ Mes), 21.2 (*p*-CH$_3$ Mes).

IR (ATR, cm^{-1}): 3013(w), 2973(w), 2917(w), 2867(w), 1572(s), 1465(s), 1302(s), 1287(s), 1251(m), 1174(m), 1055(m), 976(s), 845(s), 747(s), 693(s), 683(s), 605(m), 582(m), 554(m), 520(m).

Anal. calcd. (%) for [C$_{24}$H$_{27}$AsNK·THF] (515.61): C, 65.22; H, 6.84; N, 2.72. Found: C, 64.74; H, 6.86; N, 2.83.

4.2 Synthesis of group 4 complexes of arsinoamide

4.2.1 [(Mes$_2$AsNPh)$_2$ZrCl$_2$(THF)]·(DCM) (6)

THF (30 ml) was condensed onto ZrCl$_4$ (0.093 g, 0.4 mmol) and 2 equiv. of compound **2** (0.445 g, 0.8 mmol). After reaching room temperature, the light yellow mixture was stirred overnight at this temperature. The volatiles were removed under vacuum and DCM (20 mL) was added via cannula. The white solid was removed by filtration. The target compound could be isolated after removing DCM and washing with cold *n*-pentane. Single crystals were obtained by slow evaporation of a DCM solution. Yield: 0.246 g (58 %).

^{1}H NMR (300 MHz, d_8-THF, 298 K): δ(ppm) = 7.02-6.81 (m, 10H, Ph), 6.55 (s, 8H, ring CH Mes), 2.12 (s, 24H, *o*-CH$_3$ Mes), 1.69 (s, 12 H, *p*-CH$_3$ Mes).

^{13}C{^{1}H} NMR (75 MHz, d_8-THF, 298 K): δ(ppm) = 148.7, 143.4, 139.2, 138.3, 129.9, 129.2, 128.1, 124.4, 22.6 (*o*-CH$_3$), 20.7 (*p*-CH$_3$).

IR (ATR, cm^{-1}): 3016(w), 2965(m), 2920(m), 2862(w), 1598(m), 1584(m), 1469(s), 1445(s), 1287(m), 1202(vs), 1072(w), 1030(m), 1002(m), 915(m), 850(vs), 784(vs), 728(m), 694(vs), 676(vs), 604(m), 583(m), 548(m), 497(vs), 446(vs).

Anal. calcd. (%) for [C$_{48}$H$_{54}$As$_2$Cl$_2$N$_2$Zr·(DCM), = **6** - THF] (1055.87): C, 55.74; H, 5.35; N, 2.65. Found: C, 55.43; H, 5.22; N, 2.44.

4.2.2 [(Mes$_2$AsNPh)$_2$HfCl$_2$(THF)]·(DCM) (7)

Compound **7** was synthesized following the same synthetic approach carried out for **6** but using HfCl$_4$ instead of ZrCl$_4$.

THF (30 mL) was condensed onto HfCl$_4$ (0.128 g, 0.4 mmol) and compound **2** (0.445 g, 0.8 mmol), and the mixture was stirred overnight at room temperature. The reaction mixture turned from light yellow to colorless. All volatiles were removed under vacuum and DCM (20mL) was added via cannula. Then DCM was removed, and after washing the residue with cold *n*-pentane, the target compound was isolated. Single crystals were obtained by slow evaporation of a DCM solution. Yield: 0.208 g (47 %).

^{1}H NMR (400 MHz, d_8-THF, 298 K): δ(ppm) = 7.03-6.71 (m, 10H, Ph), 6.50 (s, 8H, ring CH Mes), 2.10 (s, 24H, *o*-CH$_3$ Mes), 1.69 (s, 12H, *p*-CH$_3$ Mes).

^{13}C{^{1}H} NMR (101 MHz, d_8-THF, 298 K): δ(ppm) = 151.1, 143.3, 141.0, 137.8, 129.8, 129.7, 127.5, 123.4, 22.8 (*o*-CH$_3$), 20.7 (*p*-CH$_3$).

IR (ATR, cm^{-1}): 3016(w), 2965(m), 2922(m), 2860(w), 1595(s), 1558(w), 1467(s), 1377(w), 1286(m), 1198(vs), 1172(m), 1072(w), 1028(m), 1000(m), 917(m), 848(vs), 789(vs), 738(m), 686(vs), 603(m), 554(m), 497(s), 448(s).

Anal. calcd. (%) for [C$_{52}$H$_{62}$As$_2$Cl$_2$HfN$_2$O, = **7** - DCM] (1130.31): C, 55.26; H, 5.53; N, 2.48. Found: C, 55.54; H, 5.23; N, 2.45.

4.2.3 [(Mes$_2$AsNPh)$_2$Zr(NMe$_2$)$_2$] (8)

Toluene (30 mL) was condensed onto (NMe$_2$)$_2$ZrCl$_2$·(THF)$_2$ (0.100 g, 0.4 mmol) and compound **2** (0.445 g, 0.8 mmol), and the mixture was stirred overnight at room temperature. After the toluene was removed under vacuum, *n*-pentane (10 mL) and diethyl ether (10 mL) were added via cannula. The white solid was removed by filtration. The filtrate was concentrated to *ca.* 1.0 mL. Single crystals were grown at -30 °C overnight. Yield: 0.206 g (52 %).

^{1}H NMR (400 MHz, C$_6$D$_6$, 298 K): δ(ppm) = 7.04-6.95 (m, 8H, *m,o*-H Ph), 6.84-6.80 (m, 2H, *p*-H Ph), 6.63 (s, 8H, ring CH Mes), 2.95 (s, 12H, NMe$_2$), 2.32 (s, 24H, *o*-CH$_3$ Mes), 2.07 (s, 12H, *p*-CH$_3$ Mes).

^{13}C{^{1}H} NMR (101 MHz, C$_6$D$_6$, 298 K): δ(ppm) = 153.8, 142.3, 138.8, 138.1, 130.1, 128.3, 126.0, 122.2, 44.0 (NMe$_2$), 22.9 (*o*-CH$_3$), 21.0 (*p*-CH$_3$).

IR (ATR, cm^{-1}): 3020(w), 2961(m), 2919(m), 2856(w), 1594(s), 1490(s), 1461(s), 1377(w), 1358(w), 1282(s), 1224(m), 1174(m), 1145(w), 1025(m), 991(w), 941(w), 903(w), 846(vs), 772(m), 747(s), 693(s), 659(s), 611(w), 582(w), 542(m), 498(m), 417(w).

Anal. calcd. (%) for [$C_{52}H_{66}As_2N_4Zr$] (988.20): C, 63.20; H, 6.73; N, 5.67. Found: C, 63.61; H, 6.88; N, 5.23.

4.2.4 [(Mes$_2$AsNPh)$_2$Hf(NMe$_2$)$_2$] (9)

Compound **9** was synthesized following the same synthetic approach carried out for **8** but using (NMe$_2$)$_2$HfCl$_2$·(THF)$_2$ instead of (NMe$_2$)$_2$ZrCl$_2$·(THF)$_2$.

Toluene (30 mL) was condensed into (NMe$_2$)$_2$HfCl$_2$·(THF)$_2$ (0.193 g, 0.4 mmol) and compound **2** (0.445 g, 0.8 mmol), and the mixture was stirred overnight at room temperature. After removing toluene under vacuum, n-pentane (10 mL) and diethyl ether (10 mL) were added via cannula. The white solid was removed by filtration. The filtrate was concentrated to *ca.* 1.0 mL. Single crystals were obtained after keeping the filtrate at -30 °C overnight. Yield: 0.142 g (33 %).

^{1}H NMR (400 MHz, C$_6$D$_6$, 298 K): δ(ppm) = 7.05-6.97 (m, 8H, *m,o*-H Ph), 6.84-6.77 (m, 2H, *p*-H Ph), 6.62 (s, 8H, ring CH Mes), 2.97 (s, 12H, NMe$_2$), 2.31 (s, 24H, *o*-CH$_3$ Mes), 2.07(s, 12H, *p*-CH$_3$ Mes).

^{13}C{^{1}H} NMR (101 MHz, C$_6$D$_6$, 298 K): δ(ppm) = 153.5, 142.3, 138.6, 138.1, 130.1, 128.1, 126.7, 122.5, 43.5 (NMe$_2$), 22.9 (*o*-CH$_3$), 20.9 (*p*-CH$_3$).

IR (ATR, cm^{-1}): 3022(w), 2961(m), 2920(m), 2855(w), 2770(w), 1595(vs), 1493(vs), 1465(s), 1445(s), 1376(m), 1357(m), 1283(vs), 1244(m), 1225(m), 1176(w), 1139(w), 1075(m), 1027(m), 993(w), 948(w), 934(w), 903(w), 870(w), 847(vs), 771(m), 748(s), 690(s), 657(m), 657(w), 611(w), 583(w), 543(m), 501(m), 416(w).

Anal. calcd. (%) for [$C_{52}H_{66}As_2HfN_4$] (1075.46): C, 58.07; H, 6.19; N, 5.21. Found: C, 58.43; H, 6.03; N, 4.77.

4.3 Synthesis of group 13 metal complexes of arsinoamide

4.3.1 [(Mes$_2$AsNPh)AlCl$_2$(THF)] (12)

THF (25 mL) was condensed at -78 °C to a mixture of compound **2** (0.889 g, 1.6 mmol) and AlCl$_3$ (0.213 g, 1.6 mmol). The mixture was allowed to warm to room temperature, and stirred at room temperature overnight. Then, the mixture was filtered and the volatiles were removed under vacuum. The residue was extracted with toluene (20 mL). The target compound was obtained after removing toluene and washing with cold n-pentane (3×5 mL). Single crystals of **12** could be obtained from slow evaporation of the dichloromethane solution. Yield: 0.52 g (57 %).

^{1}H NMR (400 MHz, C$_6$D$_6$, 298 K): δ(ppm) = 7.29-7.26 (m, 2H, m-H Ph), 7.01-6.96 (m, 2H, o-H Ph), 6.79 (s, 4H, ring CH Mes), 6.79-6.78 (m, 1H, p-H Ph), 2.62 (s, 12H, o-CH$_3$ Mes), 2.11 (s, 6H, p-CH$_3$ Mes).

^{13}C{^{1}H} NMR (101 MHz, C$_6$D$_6$, 298 K): δ(ppm) = 142.8, 138.8, 138.4, 130.5, 128.8, 125.9, 121.5, 120.4, 23.0 (o-CH$_3$), 21.0 (p-CH$_3$).

IR (ATR, cm^{-1}): 3015(w), 2959(w), 2916(w), 2863(w), 1587(m), 1481(m), 1446(m), 1229(s), 1179(w), 1081(w), 1029(m), 992(m), 956(m), 897(m), 861(s), 846(s), 805(vs), 749(m), 698(m), 610(m), 585(w), 519(vs), 473(s).

Anal. calcd. (%) for [C$_{24}$H$_{27}$AlAsCl$_2$N, = **12** -THF] (502.29): C, 57.39; H, 5.42; N, 2.79. Found: C, 57.40; H, 5.81; N, 2.50.

4.3.2 [(Mes$_2$AsNPh)InCl$_3$][Li(THF)$_4$] (**13**)

THF (25 mL) was condensed at -78 °C to a mixture of compound **2** (0.889 g, 1.6 mmol) and InCl$_3$ (0.354 g, 1.6 mmol). After stirring at room temperature overnight, the mixture was filtered and the resulting colorless solution was slowly evaporated until crystallization. The target compound was isolated after washing with cold n-pentane (3×5 mL). Yield: 0.55 g (37 %).

^{1}H NMR (400 MHz, C$_6$D$_6$, 298 K): δ(ppm) = 6.72-6.64 (m, 2H, m-H Ph), 6.64 (s, 4H, ring CH Mes), 6.32-6.30 (m, 2H, o-H Ph), 5.98-5.94 (m, 1H, p-H Ph), 2.36 (s, 12H, o-CH$_3$ Mes), 2.13 (s, 6H, p-CH$_3$ Mes).

^{13}C{^{1}H} NMR (101 MHz, C$_6$D$_6$, 298 K): δ(ppm) = 164.5, 143.2, 142.8, 136.7, 129.7, 128.7, 118.3, 109.7, 21.7 (o-CH$_3$), 20.7 (p-CH$_3$).

IR (ATR, cm^{-1}): 3016(w), 2963(w), 2918(w), 2866(w), 1588(m), 1480(m), 1446(m), 1403(w), 1378(w), 1287(w), 1229(s), 1179(w), 1080(w), 1029(m), 992(w), 956(w), 897(m), 847(s), 805(s), 748(s), 697(s), 609(s), 585(w), 518(vs), 472(vs).

Anal. calcd. (%) for [C$_{32}$H$_{43}$AsCl$_3$InLiNO$_2$, = **13** - 2 THF] (776.73): C, 49.48; H, 5.58; N, 1.80. Found: C, 49.46; H, 5.83; N, 1.58.

4.4 Synthesis of group 14 metal complexes of arsinoamide

4.4.1 [{PhC('BuN)$_2$}Si(=NPh)(AsMes$_2$)] (14)

Toluene (20 mL) was condensed onto [{PhC('BuN)$_2$}SiCl] (0.118 g, 0.4 mmol) and compound **2** (0.224 g, 0.4 mmol), and the mixture was stirred at room temperature overnight. The solid formed was removed by filtration and the target compound was isolated after removing toluene under vacuum and washing with cold *n*-pentane (3×5 mL). Single crystals were obtained from a mixture of *n*-pentane and diethyl ether (1/1) after storage at -30 °C overnight. Yield: 0.163 g (61 %).

¹H NMR (300 MHz, C$_6$D$_6$, 298 K): δ(ppm) = 7.49-7.44 (m, 2H, *m*-H Ph), 7.34-7.30 (m, 1H, *p*-H Ph), 7.23-7.20 (m, 2H, *o*-H Ph), 7.01-6.83 (m, 5H, Ph), 6.78 (s, 4H, ring CH Mes), 2.83 (s, 12H, *o*-CH$_3$ Mes), 2.09 (s, 6H, *p*-CH$_3$ Mes), 0.99 (s, 18H, *t*Bu).

¹³C{¹H} NMR (75 MHz, C$_6$D$_6$, 298 K): δ(ppm) = 176.0, 153.7, 143.6, 137.4, 136.4, 131.1, 130.6, 129.7, 129.2, 128.9, 127.9, 124.1, 115.3, 55.0, 31.1, 25.5 (*o*-CH$_3$), 20.9 (*p*-CH$_3$).

²⁹Si{¹H} NMR (59.6 MHz, C$_6$D$_6$, 298 K): δ(ppm) = -68.85.

IR (ATR, cm⁻¹): 2965(m), 2920(w), 2868(w), 1645(w), 1588(s), 1527(s), 1491(m), 1468(m), 1448(m), 1390(m), 1364(s), 1282(m), 1200(m), 1068(m), 1023(m), 920(w), 886(w), 847(s), 799(w), 697(vs), 616(m), 584(w), 552(w), 501(s), 433(w).

Anal. calcd. (%) for [C$_{39}$H$_{50}$AsN$_3$Si] (663.86): C, 70.56; H, 7.59; N, 6.33. Found: C, 70.01; H, 7.28; N, 6.18.

4.4.2 [{PhC('BuN)$_2$}Ge(Mes$_2$AsNPh)] (15)

Toluene (20 mL) was condensed onto [{PhC('BuN)$_2$}GeCl] (0.136 g, 0.4 mmol) and compound **2** (0.224 g, 0.4 mmol), and the mixture was stirred overnight at room temperature. The light yellow mixture turned to colorless. Toluene was removed under reduced pressure and *n*-pentane

(20 mL) was added via cannula. Single crystals of **15** suitable for X-ray diffraction were grown from *n*-pentane after storage at -30 °C overnight. Yield: 0.178 g (63 %).

¹H NMR (300 MHz, C₆D₆, 298 K): δ(ppm) = 7.79-7.77 (m, 2H, *m*-H Ph), 7.34-7.29 (m, 2H, *o*-H Ph), 7.22-7.20 (m, 1H, *p*-H Ph), 6.98-6.86 (m, 5H, Ph), 6.75 (s, 4H, ring CH Mes), 2.65 (s, 12H, *o*-CH₃ Mes), 2.11 (s, 6H, *p*-CH₃ Mes), 0.96 (s, 18H, *t*Bu).

¹³C{¹H} NMR (75 MHz, C₆D₆, 298 K): δ(ppm) = 167.4, 155.7, 142.5, 139.9, 137.8, 136.3, 130.2, 129.0, 128.5, 128.1, 127.2, 125.8, 120.7, 53.0, 31.8, 23.1 (*o*-CH₃), 21.0 (*p*-CH₃).

IR (ATR, cm⁻¹): 2959(m), 2920(m), 2864(w), 1587(m), 1478(m), 1458(m), 1419(vs), 1360(m), 1250(m), 1203(vs), 1177(m), 1061(m), 1017(m), 924(w), 896(w), 849(vs), 784(vs), 747(w), 722(s), 710(s), 695(vs), 604(m), 575(m), 545(m), 517(s), 479(m).

Anal. calcd. (%) for [C₃₉H₅₀AsN₃Ge] (708.40): C, 66.12; H, 7.11; N, 5.93. Found: C, 66.07; H, 6.85; N, 5.74.

4.4.3 [{Mes₂As}ClGe(μ-NPh)]₂ (**16**)

Toluene (20 mL) was condensed onto a mixture of GeCl₂·dioxane (0.093 g, 0.4 mmol) and compound **2** (0.224 g, 0.4 mmol), and the mixture was stirred for 4 h at room temperature. The solid formed was removed by filtration and the target compound was obtained after removing toluene under vacuum and washing with cold *n*-pentane (3×5 mL). Single crystals were obtained from a mixture of diethyl ether and *n*-pentane (1/1) after storage at -30 °C for 3 days. Yield: 0.074 g (36 %).

¹H NMR (300 MHz, *d*₈-THF, 298 K): δ(ppm) = 6.81-6.76 (m, 4H, *m*-H Ph), 6.73 (s, 8H, ring CH Mes), 6.66-6.61 (m, 2H, *p*-H Ph), 6.52-6.49 (m, 4H, *o*-H Ph), 2.23 (s, 24H, *o*-CH₃ Mes), 2.22 (s, 12H, *p*-CH₃ Mes).

¹³C{¹H} NMR (75 MHz, *d*₈-THF, 298 K): δ(ppm) = 153.0, 142.8, 139.2, 138.0, 130.6, 128.6, 126.0, 122.0, 23.2 (*o*-CH₃), 21.0 (*p*-CH₃).

IR (ATR, cm⁻¹): 3019(w), 2969(m), 2916(m), 2856(w), 1588(m), 1464(s), 1442(s), 1403(m), 1375(m), 1284(m), 1220(vs), 1074(w), 1024(m), 892(m), 852(vs), 778(vs), 709(s), 686(vs), 610(m), 582(m), 549(m), 503(s).

Anal. calcd. (%) for [C$_{24}$H$_{27}$AsClGeN]$_2$ (1024.98): C, 56.25; H, 5.31; N, 2.73. Found: C, 56.31; H, 5.36; N, 2.90.

4.4.4 [(Mes$_2$AsNPh)SnCl(THF)] (17)

THF (20 mL) was condensed into a mixture of SnCl$_2$ (0.152 g, 0.8 mmol) and compound **2** (0.448 g, 0.8 mmol), and the mixture was stirred overnight at room temperature. Then, THF was removed under vacuum, and toluene (15 mL) was added via cannula. The solid was removed by filtration, and compound **17** was isolated after removing toluene and washing with cold diethyl ether (3×5 mL). Single crystals were grown from a mixture of THF and diethyl ether (1/1) after storage at -30 °C for one week. Yield: 0.237 g (53 %).

^{1}H NMR (300 MHz, C$_6$D$_6$, 298 K): δ(ppm) = 7.13-7.11 (m, 2H, *m*-H Ph), 7.01-6.98 (m, 2H, *o*-H Ph), 6.84-6.80 (m, 1H, *p*-H Ph), 6.61 (s, 4H, ring CH Mes), 2.35 (s, 12H, *o*-CH$_3$ Mes), 2.05 (s, 6H, *p*-CH$_3$ Mes).

^{13}C{^{1}H} NMR (75 MHz, C$_6$D$_6$, 298 K): δ(ppm) = 155.6, 143.3, 139.0, 136.6, 130.5, 129.2, 122.3, 120.1, 22.6 (*o*-CH$_3$), 20.9 (*p*-CH$_3$).

^{119}Sn{^{1}H} NMR (112 MHz, C$_6$D$_6$, 298 K): δ(ppm) = -34.2.

IR (ATR, cm^{-1}): 3021(w), 2964(m), 2920(m), 2868(w), 1595(s), 1491(s), 1462(s), 1377(w), 1358(w), 1282(vs), 1173(m), 1043(m), 1030(m), 907(w), 847(s), 747(s), 690(s), 589(m), 554(m), 499(m), 443(m).

Anal. calcd. (%) for [C$_{24}$H$_{27}$AsClNSn = **17** - THF] (558.57): C, 51.61; H, 4.87; N, 2.51. Found: C, 51.68; H, 5.29; N, 2.08.

4.4.5 [(Mes$_2$AsNPh)$_2$Ge] (18)

Toluene (20 mL) was condensed onto a mixture of GeCl$_2$·dioxane (0.093 g, 0.4 mmol) and compound **2** (0.448 g, 0.8 mmol), and the mixture was stirred 3.0 h at room temperature. The solid formed was removed by filtration and the compound **18** was obtained as a yellow powder after removing toluene under vacuum and washing with cold *n*-pentane. Single crystals of **18** were crystallized from *n*-heptane at -30 °C overnight. Yield: 0.307 g (87 %).

¹H NMR (300 MHz, C₆D₆, 298 K): δ(ppm) = 6.99-6.94 (m, 4H, Ph), 6.86-6.77 (m, 6H, Ph), 6.63 (s, 8H, ring CH Mes), 2.39 (s, 24H, *o*-CH₃ Mes), 2.07 (s, 12H, *p*-CH₃ Mes).

¹³C{¹H} NMR (75 MHz, C₆D₆, 298 K): δ(ppm) = 152.8, 142.4, 138.6, 137.7, 130.4, 128.5, 125.7, 121.9, 23.2 (*o*-CH₃), 20.9 (*p*-CH₃).

IR (ATR, cm⁻¹): 3019(w), 2970(w), 2948(w), 2916(w), 1586(m), 1463(m), 1442(m), 1375(w), 1285(w), 1221(s), 1075(w), 1026(m), 892(m), 850(s), 779(vs), 709(s), 687(vs), 612(w), 581(w), 547(w), 504(s).

Anal. calcd. (%) for [C₄₈H₅₄As₂GeN₂] (881.45): C, 65.41; H, 6.18; N, 3.18. Found: C, 65.65; H, 6.07; N, 3.12.

4.4.6 [(Mes₂AsNPh)₂Sn] (**19**)

Toluene (20 mL) was condensed onto SnCl₂ (0.076 g, 0.4 mmol) and compound **2** (0.448 g, 0.8 mmol), and the mixture was stirred at room temperature for 6 h. The solid formed was removed by filtration and compound **19** was isolated as a red solid after removing toluene under vacuum and washing with cold *n*-pentane. Single crystals were crystallized by diffusion of *n*-pentane and diethyl ether (1/1) into a solution of **19** in toluene. Yield: 0.267 g (72 %).

¹H NMR (300 MHz, C₆D₆, 298 K): δ(ppm) = 7.14-7.11 (m, 4H, *m*-H Ph), 7.01-6.98 (m, 4H, *o*-H Ph), 6.84-6.80 (m, 2H, *p*-H Ph), 6.61 (s, 8H, ring CH Mes), 2.35 (s, 24H, *o*-CH₃ Mes), 2.05 (s, 12H, *p*-CH₃ Mes).

¹³C{¹H} NMR (75 MHz, C₆D₆, 298 K): δ(ppm) = 155.6, 143.3, 139.0, 136.6, 130.5, 129.2, 122.3, 120.1, 22.6 (*o*-CH₃), 20.9 (*p*-CH₃).

¹¹⁹Sn{¹H} NMR (112 MHz, C₆D₆, 298 K): δ(ppm) = 318.6.

IR (ATR, cm⁻¹): 3017(w), 2961(w), 2917(w), 2854(w), 1595(vs), 1493(vs), 1465(s), 1442(s), 1375(m), 1356(m), 1283(vs), 1227(s), 1174(m), 1075(m), 1026(m), 848(s), 760(vs), 749(s), 684(vs), 611(w), 583(w), 558(w), 543(m), 512(m), 463(w).

Anal. calcd. (%) for [C₄₈H₅₄As₂N₂Sn] (927.53): C, 62.16; H, 5.87; N, 3.02. Found: C, 62.51; H, 5.56; N, 2.65.

4.4.7 [(Mes$_2$AsNPh)$_2$Pb] (**20**)

Toluene (20 mL) was condensed into PbCl$_2$ (0.111 g, 0.4 mmol) and compound **2** (0.448 g, 0.8 mmol), and the mixture was stirred overnight at room temperature. The solid formed was removed by filtration and compound **20** was obtained as a deep violated powder after removing toluene and washing with cold *n*-pentane (2×5 mL) and diethyl ether (5 mL). Single crystals were obtained by diffusion of a mixture of *n*-pentane and diethyl ether (1/1) into a solution of **20** in toluene. Yield: 0.229 g (56 %).

^{1}H NMR (300 MHz, C$_6$D$_6$, 298 K): δ(ppm) = 7.36-7.31 (m, 4H, *m*-H Ph), 7.13-7.10 (m, 4H, *o*-H Ph), 6.78-6.71 (m, 2H, *p*-H Ph), 6.60 (s, 8H, ring CH Mes), 2.34 (s, 24H, *o*-CH$_3$ Mes), 2.04 (s, 12H, *p*-CH$_3$ Mes).

^{13}C{^{1}H} NMR (75 MHz, C$_6$D$_6$, 298 K): δ(ppm) = 157.7, 143.6, 139.0, 137.0, 130.3, 129.3, 121.4, 119.1, 22.1 (*o*-CH$_3$), 20.9 (*p*-CH$_3$).

^{207}Pb{^{1}H} NMR (62.8MHz, C$_6$D$_6$, 298 K): δ(ppm) = 3244.

IR (ATR, cm^{-1}): 3021(w), 2958(m), 2918(m), 2866(w), 1595(vs), 1492(vs), 1465(s), 1443(s), 1357(m), 1283(vs), 1228(m), 1152(m), 1027(m), 847(s), 750(s), 691(s), 543(m), 507(s).

Anal. calcd. (%) for [C$_{48}$H$_{54}$As$_2$N$_2$Pb] (1016.02): C, 56.74; H, 5.36; N, 2.76. Found: C, 57.18; H, 5.34; N, 2.83.

4.4.8 [(Mes$_2$AsNPh)$_2$GeNi(COD)] (**21**)

To a stirring solution of [(PhNAsMes$_2$)$_2$Ge] (0.353 g, 0.4 mmol) in 20 mL diethyl ether, a solution of Ni(COD)$_2$ (0.110 g, 0.4 mmol) in 40 mL diethyl ether was added dropwise. The mixture was stirred for 15 min at room temperature, leading to the formation of micro crystals. After keep the mixture under -30 °C for one week, single crystals suitable for X-ray diffraction studies were obtained. Compound **21** was isolated as a red-brown solid after filtration and washing with cold *n*-pentane (2×5 mL). Yield: 0.098 g (23 %).

IR (ATR, cm^{-1}): 2963(m), 2920(m), 2866(w), 1645(m), 1593(s), 1528(m), 1488(vs), 1446(s), 1387(m), 1362(m), 1285(s), 1224(s), 1204(s), 1073(m), 1026(m), 919(m), 882(m), 848(s), 775(m), 748(m), 697(s), 616(w), 584(w), 552(w), 502(m).

Anal. calcd. (%) for [C$_{56}$H$_{66}$As$_2$GeN$_2$Ni] (1048.32): C, 64.16; H, 6.35; N, 2.67. Found: C, 63.77; H, 5.87; N, 2.86.

5 Crystal Data

5.1 Data collection and refinement

The suitable crystals were covered in mineral oil (Aldrich) and mounted on a glass fiber or a mylar loop. The crystal was transferred directly to the cold stream of a STOE IPDS 2 (150 or 210 K) or STADIVARI (100 K) diffractometer.

All structures were solved by using the program SHELXS/T[212,213] using Olex2[214] and refined with the ShelXL[215] refinement package using Least Squares minimization. The remaining non-hydrogen atoms were located from successive difference Fourier map calculations. The refinements were carried out by using full-matrix least-squares techniques on F^2 by using the program SHELXL. In each case, the locations of the largest peaks in the final difference Fourier map calculations, as well as the magnitude of the residual electron densities, were of no chemical significance. The molecule structures of compounds **2-21** were given by Diamond 4.5.2.[216]

For compound **2**, one of the diethyl ether molecules position is about 40% occupied by a THF molecule.

For compound **11**, **17** and **21**, owing to the blemish of crystal, only low quality data was collected.

For compound **19** and **20**, disordered n-pentane molecules were removed using the Olex2 solvent mask command.

5.2 Crystal data

5.2.1 [(Mes$_2$AsNPh){Li(OEt$_2$)$_2$}] (2)

Compound	2
Empirical formula	0.6(C$_{32}$H$_{47}$AsLiNO$_2$)·0.4(C$_{32}$H$_{45}$AsLiNO$_2$)
Formula weight	558.76
Temperature/K	150
Crystal system	monoclinic
Space group	$P2_1$
a/Å	9.763(2)
b/Å	16.347(3)
c/Å	10.442(2)
β/°	109.81(3)
Volume/Å^3	1567.9(6)
Z	2
ρ_{calc}g/cm^3	1.185
μ/mm^{-1}	1.110
F(000)	596.0
Crystal size/mm^3	0.357 × 0.286 × 0.159
Radiation	MoKα (λ = 0.71073)
2Θ range for data collection/°	4.146 to 52.182
Index ranges	-11 ≤ h ≤ 12, -20 ≤ k ≤ 20, -12 ≤ l ≤ 12
Reflections collected	26044
Independent reflections	6174 [R$_{int}$ = 0.0658, R$_{sigma}$ = 0.0502]
Data/restraints/parameters	6174/1/344
Goodness-of-fit on F^2	1.005
Final R indexes [I>=2σ (I)]	R$_1$ = 0.0636, wR$_2$ = 0.1621
Final R indexes [all data]	R$_1$ = 0.0744, wR$_2$ = 0.1682
Largest diff. peak/hole / e Å^{-3}	0.88/-0.75
Flack parameter	-0.02(2)

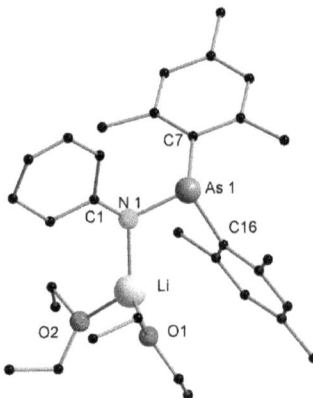

5.2.2 [(Mes₂AsNPh){Na(OEt₂)}]₂ (3)

Compound	3
Empirical formula	$C_{28}H_{37}AsNNaO$
Formula weight	501.49
Temperature/K	200
Crystal system	triclinic
Space group	$P\text{-}1$
a/Å	10.405(2)
b/Å	10.549(2)
c/Å	13.329(3)
α/°	82.21(3)
β/°	75.57(3)
γ/°	69.79(3)
Volume/Å³	1327.7(6)
Z	2
ρ_{calc}g/cm³	1.254
μ/mm⁻¹	1.316
F(000)	528.0
Crystal size/mm³	0.326 × 0.229 × 0.141
Radiation	MoKα (λ = 0.71073)
2Θ range for data collection/°	3.16 to 52.024
Index ranges	$-12 \le h \le 11, -12 \le k \le 13, -16 \le l \le 16$
Reflections collected	9162
Independent reflections	4998 [R_{int} = 0.0597, R_{sigma} = 0.0976]
Data/restraints/parameters	4998/0/297
Goodness-of-fit on F^2	0.896
Final R indexes [I>=2σ (I)]	R_1 = 0.0553, wR_2 = 0.1229
Final R indexes [all data]	R_1 = 0.0875, wR_2 = 0.1328
Largest diff. peak/hole / e Å⁻³	0.91/-0.83

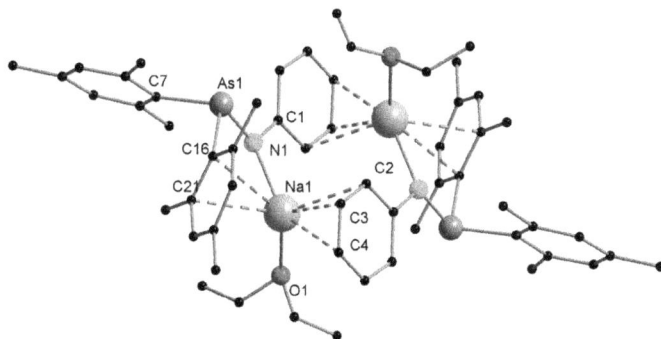

5.2.3 [(Mes₂AsNPh){K(THF)}]₂ (4)

Compound	4
Empirical formula	$C_{28}H_{35}AsKNO$
Formula weight	515.59
Temperature/K	200
Crystal system	triclinic
Space group	P-1
a/Å	9.736(2)
b/Å	10.771(2)
c/Å	14.012(3)
$\alpha/°$	67.46(3)
$\beta/°$	82.00(3)
$\gamma/°$	76.13(3)
Volume/Å³	1315.6(6)
Z	1
ρ_{calc}g/cm³	1.302
μ/mm⁻¹	1.469
F(000)	540.0
Crystal size/mm³	0.339 × 0.255 × 0.111
Radiation	MoKα (λ = 0.71073)
2Θ range for data collection/°	3.152 to 52
Index ranges	$-11 \leq h \leq 11, -11 \leq k \leq 13, -17 \leq l \leq 17$
Reflections collected	10197
Independent reflections	5129 [R_{int} = 0.0867, R_{sigma} = 0.0897]
Data/restraints/parameters	5129/0/295
Goodness-of-fit on F²	0.938
Final R indexes [I>=2σ (I)]	R_1 = 0.0528, wR_2 = 0.1253
Final R indexes [all data]	R_1 = 0.0778, wR_2 = 0.1353
Largest diff. peak/hole / e Å⁻³	0.49/-0.39

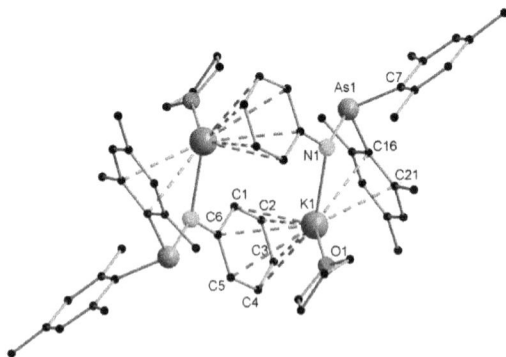

5.2.4 [(Mes₂AsNPh){Li(2,2'-bpy)}] (5)

5.2.4 [(Mes$_2$AsNPh){Li(2,2'-bpy)}] (**5**)

Compound	5
Empirical formula	$C_{41}H_{43}AsLiN_3$
Formula weight	659.64
Temperature/K	150
Crystal system	triclinic
Space group	P-1
a/Å	13.503(3)
b/Å	15.184(3)
c/Å	18.007(4)
α/°	105.41(3)
β/°	99.08(3)
γ/°	94.56(3)
Volume/Å³	3485.9(14)
Z	4
ρ_{calc}g/cm³	1.257
μ/mm⁻¹	1.007
F(000)	1384.0
Crystal size/mm³	0.231 × 0.167 × 0.081
Radiation	MoKα (λ = 0.71073)
2Θ range for data collection/°	3.08 to 58.528
Index ranges	-18 ≤ h ≤ 17, -20 ≤ k ≤ 20, -24 ≤ l ≤ 23
Reflections collected	35134
Independent reflections	18679 [R_{int} = 0.0697, R_{sigma} = 0.1242]
Data/restraints/parameters	18679/0/843
Goodness-of-fit on F²	1.024
Final R indexes [I>=2σ (I)]	R_1 = 0.0655, wR_2 = 0.1255
Final R indexes [all data]	R_1 = 0.1516, wR_2 = 0.1540
Largest diff. peak/hole / e Å⁻³	1.15/-0.53

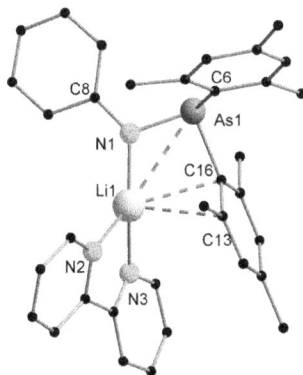

5.2.5 [(Mes$_2$AsNPh)$_2$ZrCl$_2$(THF)]·(DCM) (6)

Compound	6
Empirical formula	C$_{53}$H$_{64}$As$_2$Cl$_4$N$_2$OZr
Formula weight	1127.92
Temperature/K	100
Crystal system	monoclinic
Space group	$P2_1/n$
a/Å	12.370(3)
b/Å	16.883(3)
c/Å	25.027(5)
β/°	102.37(3)
Volume/Å^3	5105.2(2)
Z	4
ρ_{calc}g/cm^3	1.467
μ/mm^{-1}	1.751
F(000)	2312.0
Crystal size/mm^3	0.162 × 0.131 × 0.077
Radiation	MoKα (λ = 0.71073)
2Θ range for data collection/°	4.068 to 52.358
Index ranges	-15 ≤ h ≤ 15, -20 ≤ k ≤ 20, -15 ≤ l ≤ 30
Reflections collected	20263
Independent reflections	9989 [R$_{int}$ = 0.0434, R$_{sigma}$ = 0.0661]
Data/restraints/parameters	9989/0/580
Goodness-of-fit on F^2	1.014
Final R indexes [I>=2σ (I)]	R$_1$ = 0.0492, wR$_2$ = 0.1037
Final R indexes [all data]	R$_1$ = 0.0843, wR$_2$ = 0.1194
Largest diff. peak/hole / e Å^{-3}	0.68/-0.54

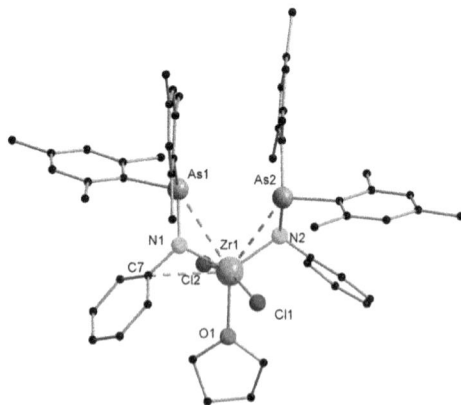

5.2.6 [(Mes$_2$AsNPh)$_2$HfCl$_2$(THF)]·(DCM) (7)

Compound	7
Empirical formula	C$_{53}$H$_{64}$As$_2$Cl$_4$HfN$_2$O
Formula weight	1215.19
Temperature/K	100.0
Crystal system	monoclinic
Space group	$P2_1/n$
a/Å	12.3569(6)
b/Å	16.8984(7)
c/Å	24.8863(13)
β/°	102.510(4)
Volume/Å^3	5073.2(4)
Z	4
ρ_{calc}g/cm^3	1.591
μ/mm^{-1}	3.601
F(000)	2440.0
Crystal size/mm^3	0.404 × 0.211 × 0.078
Radiation	MoKα (λ = 0.71073)
2Θ range for data collection/°	4.822 to 52.288
Index ranges	-14 ≤ h ≤ 15, -20 ≤ k ≤ 20, -30 ≤ l ≤ 27
Reflections collected	24740
Independent reflections	9996 [R$_{int}$ = 0.0300, R$_{sigma}$ = 0.0349]
Data/restraints/parameters	9996/0/580
Goodness-of-fit on F^2	1.023
Final R indexes [I>=2σ (I)]	R$_1$ = 0.0334, wR$_2$ = 0.0825
Final R indexes [all data]	R$_1$ = 0.0416, wR$_2$ = 0.0872
Largest diff. peak/hole / e Å^{-3}	1.42/-0.67

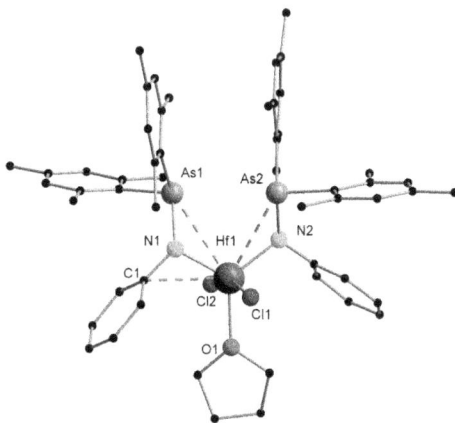

5.2.7 [(Mes$_2$AsNPh)$_2$Zr(NMe$_2$)$_2$] (8)

Compound	8
Empirical formula	C$_{52}$H$_{66}$As$_2$N$_4$Zr
Formula weight	988.14
Temperature/K	100
Crystal system	triclinic
Space group	P-1
a/Å	9.3167(7)
b/Å	13.5127(10)
c/Å	19.5543(14)
α/°	95.895(6)
β/°	95.785(6)
γ/°	103.126(6)
Volume/Å^3	2365.1(3)
Z	2
ρ_{calc}g/cm^3	1.388
μ/mm^{-1}	1.660
F(000)	1024.0
Crystal size/mm^3	0.373 × 0.336 × 0.314
Radiation	MoKα (λ = 0.71073)
2Θ range for data collection/°	3.534 to 63.208
Index ranges	-13 ≤ h ≤ 13, -19 ≤ k ≤ 18, -28 ≤ l ≤ 27
Reflections collected	23999
Independent reflections	12895 [R$_{int}$ = 0.0347, R$_{sigma}$ = 0.0446]
Data/restraints/parameters	12895/0/548
Goodness-of-fit on F^2	1.054
Final R indexes [I>=2σ (I)]	R$_1$ = 0.0392, wR$_2$ = 0.0985
Final R indexes [all data]	R$_1$ = 0.0526, wR$_2$ = 0.1052
Largest diff. peak/hole / e Å^{-3}	0.72/-0.79

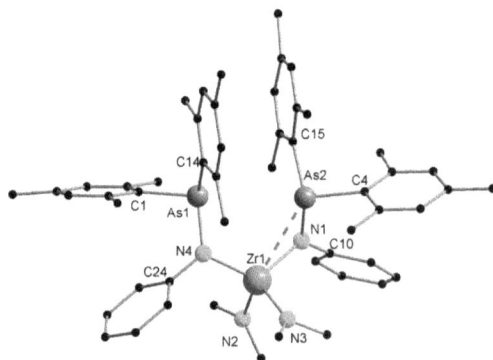

5.2.8 [(Mes$_2$AsNPh)$_2$Hf(NMe$_2$)$_2$] (9)

Compound	9
Empirical formula	C$_{52}$H$_{66}$As$_2$HfN$_4$
Formula weight	1075.41
Temperature/K	150
Crystal system	triclinic
Space group	*P*-1
a/Å	9.3333(6)
b/Å	13.5745(7)
c/Å	19.6722(11)
α/°	95.768(4)
β/°	95.975(5)
γ/°	103.193(5)
Volume/Å^3	2393.4(2)
Z	2
ρ$_{calc}$g/cm^3	1.492
μ/mm^{-1}	3.590
F(000)	1088.0
Crystal size/mm^3	0.438 × 0.311 × 0.182
Radiation	MoKα (λ = 0.71073)
2Θ range for data collection/°	3.52 to 59
Index ranges	-12 ≤ h ≤ 12, -18 ≤ k ≤ 16, -26 ≤ l ≤ 27
Reflections collected	26669
Independent reflections	13195 [R$_{int}$ = 0.0290, R$_{sigma}$ = 0.0396]
Data/restraints/parameters	13195/0/548
Goodness-of-fit on F^2	1.031
Final R indexes [I>=2σ (I)]	R$_1$ = 0.0283, wR$_2$ = 0.0644
Final R indexes [all data]	R$_1$ = 0.0407, wR$_2$ = 0.0675
Largest diff. peak/hole / e Å^{-3}	1.74/-1.34

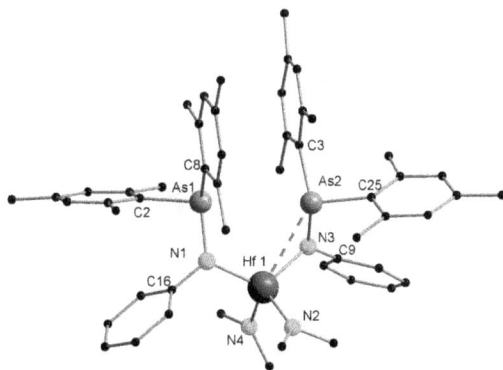

5.2.9 [(Mes$_2$AsNPh){MgBr(THF)}]$_2$ (10)

Compound	10
Empirical formula	C$_{56}$H$_{70}$As$_2$Br$_2$Mg$_2$N$_2$O$_2$
Formula weight	1161.42
Temperature/K	100.0
Crystal system	monoclinic
Space group	$P2_1/n$
a/Å	10.8278(6)
b/Å	21.8075(2)
c/Å	11.0738(5)
β/°	93.585(4)
Volume/Å^3	2609.7(3)
Z	2
ρ_{calc}g/cm^3	1.478
μ/mm^{-1}	2.879
F(000)	1192.0
Crystal size/mm^3	0.119 × 0.09 × 0.048
Radiation	MoKα (λ = 0.71073)
2Θ range for data collection/°	3.736 to 51.988
Index ranges	-12 ≤ h ≤ 13, -26 ≤ k ≤ 26, -13 ≤ l ≤ 13
Reflections collected	24084
Independent reflections	5133 [R$_{int}$ = 0.0872, R$_{sigma}$ = 0.0522]
Data/restraints/parameters	5133/0/304
Goodness-of-fit on F^2	1.046
Final R indexes [I>=2σ (I)]	R$_1$ = 0.0599, wR$_2$ = 0.1686
Final R indexes [all data]	R$_1$ = 0.0766, wR$_2$ = 0.1805
Largest diff. peak/hole / e Å^{-3}	1.39/-2.06

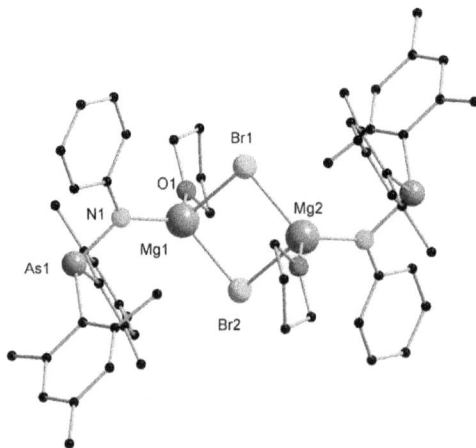

5.2.10 [(Mes₂As)₂NPh] (11)

Compound	11
Empirical formula	$C_{42}H_{49}As_2N$
Formula weight	717.66
Temperature/K	210
Crystal system	triclinic
Space group	P-1
a/Å	9.0214(7)
b/Å	13.5984(12)
c/Å	15.6399(2)
α/°	108.828(8)
β/°	94.135(8)
γ/°	97.203(7)
Volume/Å^3	1788.6(3)
Z	2
ρ_{calc}g/cm^3	1.333
μ/mm^{-1}	1.898
F(000)	748.0
Crystal size/mm^3	0.202 × 0.184 × 0.155
Radiation	MoKα ($\lambda = 0.71073$)
2Θ range for data collection/°	3.204 to 52.036
Index ranges	$-11 \leq h \leq 11, -16 \leq k \leq 16, -17 \leq l \leq 19$
Reflections collected	13900
Independent reflections	6967 [$R_{int} = 0.1114$, $R_{sigma} = 0.1509$]
Data/restraints/parameters	6967/0/418
Goodness-of-fit on F^2	0.957
Final R indexes [I>=2σ (I)]	$R_1 = 0.0776$, $wR_2 = 0.1940$
Final R indexes [all data]	$R_1 = 0.1469$, $wR_2 = 0.2192$
Largest diff. peak/hole / e Å^{-3}	1.15/-0.58

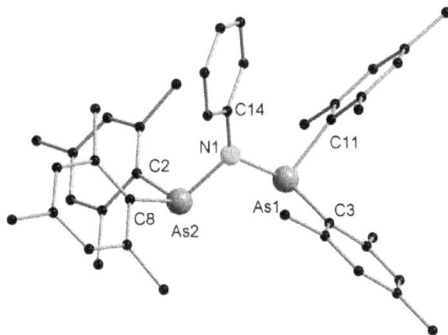

5.2.11 [(Mes₂AsNPh)AlCl₂(THF)] (12)

Compound	12
Empirical formula	$C_{28}H_{35}AlAsCl_2NO$
Formula weight	572.35
Temperature/K	210
Crystal system	triclinic
Space group	$P\text{-}1$
a/Å	8.309(2)
b/Å	12.830(3)
c/Å	13.327(3)
α/°	86.65(3)
β/°	87.32(3)
γ/°	85.59(3)
Volume/Å³	1412.9(5)
Z	2
ρ_{calc}g/cm³	1.345
μ/mm⁻¹	1.444
F(000)	592.0
Crystal size/mm³	0.259 × 0.206 × 0.121
Radiation	MoKα (λ = 0.71073)
2Θ range for data collection/°	3.188 to 59.098
Index ranges	$-9 \leq h \leq 11, -17 \leq k \leq 17, -18 \leq l \leq 18$
Reflections collected	14558
Independent reflections	7780 [R_{int} = 0.0433, R_{sigma} = 0.0788]
Data/restraints/parameters	7780/0/313
Goodness-of-fit on F^2	1.034
Final R indexes [I>=2σ (I)]	R_1 = 0.0638, wR_2 = 0.1253
Final R indexes [all data]	R_1 = 0.1071, wR_2 = 0.1402
Largest diff. peak/hole / e Å⁻³	0.64/-0.38

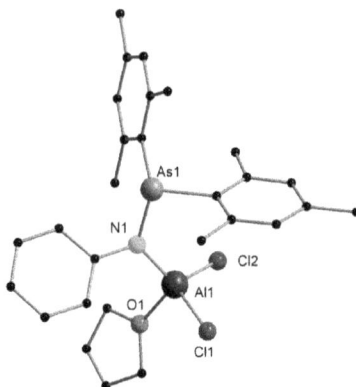

5.2.12 [(Mes₂AsNPh)InCl₃][Li(THF)₄] (**13**)

Compound	13
Empirical formula	$C_{40}H_{59}AsCl_3InLiNO_4$
Formula weight	920.91
Temperature/K	100.0
Crystal system	monoclinic
Space group	$P2_1/n$
a/Å	15.2723(8)
b/Å	13.4957(5)
c/Å	20.3619(11)
β/°	95.307(4)
Volume/Å³	4178.8(4)
Z	4
ρ_{calc}g/cm³	1.464
μ/mm⁻¹	1.583
F(000)	1896.0
Crystal size/mm³	0.409 × 0.317 × 0.22
Radiation	MoKα (λ = 0.71073)
2Θ range for data collection/°	3.494 to 51.996
Index ranges	$-17 \leq h \leq 18, -16 \leq k \leq 14, -19 \leq l \leq 25$
Reflections collected	22916
Independent reflections	8169 [R_{int} = 0.0673, R_{sigma} = 0.0439]
Data/restraints/parameters	8169/0/466
Goodness-of-fit on F²	1.110
Final R indexes [I>=2σ (I)]	R_1 = 0.0322, wR_2 = 0.0787
Final R indexes [all data]	R_1 = 0.0453, wR_2 = 0.0899
Largest diff. peak/hole / e Å⁻³	0.53/-0.62

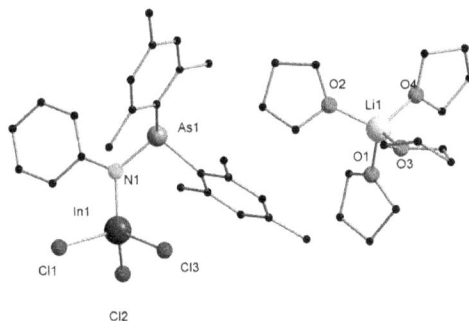

5.2.13 [{PhC(^{t}BuN)$_2$}Si(=NPh)(AsMes$_2$)] (14)

Compound	14
Empirical formula	C$_{39}$H$_{50}$AsN$_3$Si
Formula weight	663.83
Temperature/K	150.0
Crystal system	monoclinic
Space group	$P2_1/c$
a/Å	16.904(2)
b/Å	12.7862(10)
c/Å	18.468(2)
β/°	116.255(9)
Volume/Å^3	3579.8(7)
Z	4
ρ$_{calc}$g/cm^3	1.232
μ/mm^{-1}	1.013
F(000)	1408.0
Crystal size/mm^3	0.393 × 0.324 × 0.213
Radiation	MoKα (λ = 0.71073)
2Θ range for data collection/°	4.024 to 58.934
Index ranges	-23 ≤ h ≤ 23, -17 ≤ k ≤ 15, -24 ≤ l ≤ 25
Reflections collected	26585
Independent reflections	9915 [R$_{int}$ = 0.0330, R$_{sigma}$ = 0.0317]
Data/restraints/parameters	9915/0/409
Goodness-of-fit on F^2	1.084
Final R indexes [I>=2σ (I)]	R$_1$ = 0.0449, wR$_2$ = 0.1089
Final R indexes [all data]	R$_1$ = 0.0640, wR$_2$ = 0.1254
Largest diff. peak/hole / e Å^{-3}	0.77/-0.62

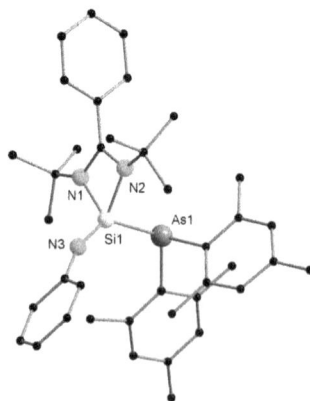

5.2.14 [{PhC('BuN)₂}Ge(Mes₂AsNPh)] (15)

Compound	15
Empirical formula	$C_{39}H_{50}AsGeN_3$
Formula weight	708.33
Temperature/K	150.0
Crystal system	triclinic
Space group	$P\text{-}1$
a/Å	9.1960(11)
b/Å	12.0628(12)
c/Å	16.232(2)
α/°	91.405(10)
β/°	90.933(10)
γ/°	93.901(9)
Volume/Å³	1795.6(4)
Z	2
ρ_{calc}g/cm³	1.310
μ/mm⁻¹	1.798
F(000)	740.0
Crystal size/mm³	0.48 × 0.378 × 0.269
Radiation	MoKα (λ = 0.71073)
2Θ range for data collection/°	3.386 to 52.008
Index ranges	-11 ≤ h ≤ 11, -14 ≤ k ≤ 14, -19 ≤ l ≤ 18
Reflections collected	14024
Independent reflections	6990 [R_{int} = 0.0534, R_{sigma} = 0.0510]
Data/restraints/parameters	6990/0/409
Goodness-of-fit on F^2	0.987
Final R indexes [I>=2σ (I)]	R_1 = 0.0323, wR_2 = 0.0809
Final R indexes [all data]	R_1 = 0.0469, wR_2 = 0.0856
Largest diff. peak/hole / e Å⁻³	0.73/-0.73

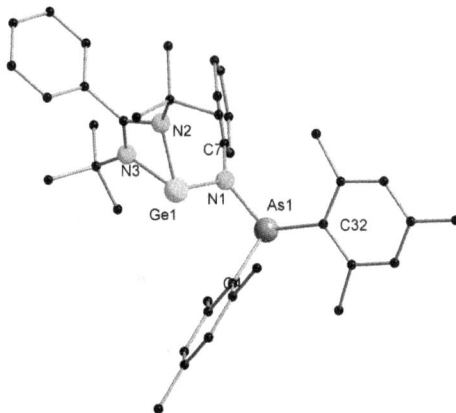

5.2.15 [{Mes$_2$As}ClGe(μ-NPh)]$_2$ (16)

Compound	16
Empirical formula	$C_{48}H_{54}As_2Cl_2Ge_2N_2$
Formula weight	1024.85
Temperature/K	100
Crystal system	monoclinic
Space group	$P2_1/c$
a/Å	12.368(3)
b/Å	8.544(2)
c/Å	24.172(5)
$\beta/°$	98.89(3)
Volume/Å^3	2523.7(9)
Z	2
ρ_{calc}g/cm^3	1.349
μ/mm^{-1}	2.630
F(000)	1040.0
Crystal size/mm^3	0.174 × 0.09 × 0.044
Radiation	MoKα (λ = 0.71073)
2Θ range for data collection/°	5.064 to 59.302
Index ranges	-16 ≤ h ≤ 16, -11 ≤ k ≤ 10, -32 ≤ l ≤ 33
Reflections collected	14014
Independent reflections	6184 [R_{int} = 0.0491, R_{sigma} = 0.0729]
Data/restraints/parameters	6184/0/259
Goodness-of-fit on F^2	1.035
Final R indexes [I>=2σ (I)]	R_1 = 0.0595, wR_2 = 0.1304
Final R indexes [all data]	R_1 = 0.1067, wR_2 = 0.1567
Largest diff. peak/hole / e Å^{-3}	1.16/-1.12

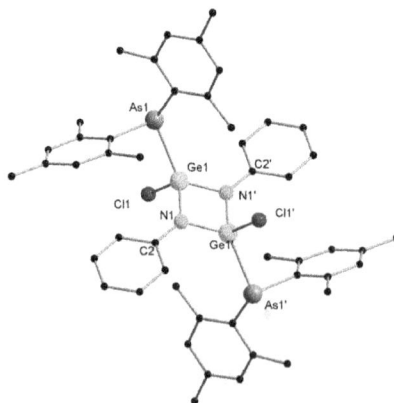

5.2.16 [(Mes₂AsNPh)SnCl(THF)] (17)

5.2.16 [(Mes$_2$AsNPh)SnCl(THF)] (**17**)

Compound	17
Empirical formula	$C_{28}H_{35}AsClNOSn$
Formula weight	686.10
Temperature/K	100
Crystal system	triclinic
Space group	P-1
a/Å	8.086(2)
b/Å	12.682(3)
c/Å	13.284(3)
α/°	87.88(3)
β/°	88.19(3)
γ/°	85.19(3)
Volume/Å^3	1355.9(5)
Z	2
ρ_{calc}g/cm^3	1.680
μ/mm^{-1}	2.652
F(000)	684.0
Crystal size/mm^3	0.203 × 0.172 × 0.143
Radiation	MoKα (λ = 0.71073)
2Θ range for data collection/°	4.528 to 61.702
Index ranges	-11 ≤ h ≤ 10, -17 ≤ k ≤ 17, -18 ≤ l ≤ 16
Reflections collected	12510
Independent reflections	6688 [R_{int} = 0.0376, R_{sigma} = 0.0383]
Data/restraints/parameters	6688/0/304
Goodness-of-fit on F^2	1.114
Final R indexes [I>=2σ (I)]	R_1 = 0.0487, wR_2 = 0.1332
Final R indexes [all data]	R_1 = 0.0561, wR_2 = 0.1431
Largest diff. peak/hole / e Å^{-3}	2.68/-1.40

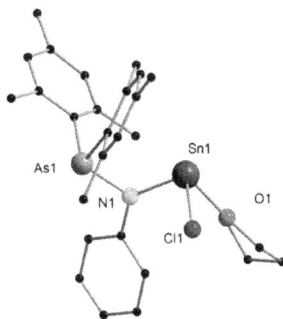

5.2.17 [(Mes$_2$AsNPh)$_2$Ge] (18)

Compound	18
Empirical formula	C$_{48}$H$_{54}$As$_2$GeN$_2$
Formula weight	881.36
Temperature/K	210
Crystal system	monoclinic
Space group	C2/c
a/Å	13.018(3)
b/Å	18.028(4)
c/Å	19.158(4)
β/°	107.77(3)
Volume/Å^3	4281.8(2)
Z	4
ρ$_{calc}$g/cm^3	1.367
μ/mm^{-1}	2.284
F(000)	1816.0
Crystal size/mm^3	0.204 × 0.151 × 0.073
Radiation	MoKα (λ = 0.71073)
2Θ range for data collection/°	3.988 to 58.98
Index ranges	-18 ≤ h ≤ 17, -22 ≤ k ≤ 24, -25 ≤ l ≤ 26
Reflections collected	15103
Independent reflections	5950 [R$_{int}$ = 0.0214, R$_{sigma}$ = 0.0342]
Data/restraints/parameters	5950/0/246
Goodness-of-fit on F^2	0.954
Final R indexes [I>=2σ (I)]	R$_1$ = 0.0314, wR$_2$ = 0.0636
Final R indexes [all data]	R$_1$ = 0.0522, wR$_2$ = 0.0685
Largest diff. peak/hole / e Å^{-3}	0.42/-0.23

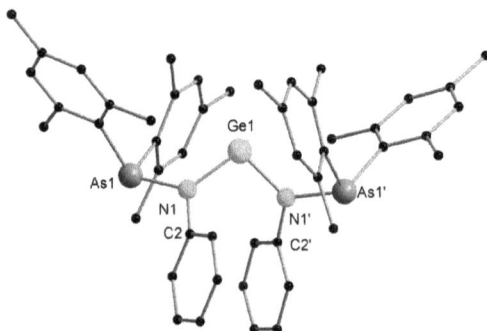

5.2.18 [(Mes₂AsNPh)₂Sn] (19)

Compound	19
Empirical formula	$C_{48}H_{54}As_2N_2Sn$
Formula weight	977.62
Temperature/K	100
Crystal system	tetragonal
Space group	$I4_1/a$
a/Å	23.658(3)
c/Å	31.714(6)
Volume/Å^3	17751(6)
Z	16
ρ_{calc}g/cm^3	1.448
μ/mm^{-1}	2.093
F(000)	7869.0
Crystal size/mm^3	0.4 × 0.179 × 0.067
Radiation	MoKα (λ = 0.71073)
2Θ range for data collection/°	3.444 to 49.992
Index ranges	-28 ≤ h ≤ 28, -28 ≤ k ≤ 28, -37 ≤ l ≤ 37
Reflections collected	77069
Independent reflections	7824 [R_{int} = 0.2478, R_{sigma} = 0.0790]
Data/restraints/parameters	7824/0/532
Goodness-of-fit on F^2	1.056
Final R indexes [I>=2σ (I)]	R_1 = 0.0609, wR_2 = 0.1465
Final R indexes [all data]	R_1 = 0.0773, wR_2 = 0.1640
Largest diff. peak/hole / e Å^{-3}	1.71/-1.74

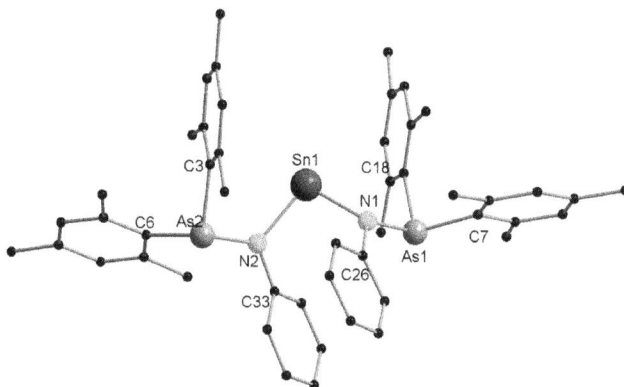

5.2.19 [(Mes₂AsNPh)₂Pb] (20)

Compound	20
Empirical formula	$C_{48}H_{54}As_2N_2Pb$
Formula weight	1015.96
Temperature/K	100
Crystal system	tetragonal
Space group	$I4_1/a$
a/Å	23.697(3)
c/Å	31.548(6)
Volume/Å^3	17715(6)
Z	16
ρ_{calc}g/cm^3	1.524
μ/mm^{-1}	5.326
F(000)	8064.0
Crystal size/mm^3	0.386 × 0.159 × 0.041
Radiation	MoKα (λ = 0.71073)
2Θ range for data collection/°	4.862 to 62.15
Index ranges	-30 ≤ h ≤ 31, -31 ≤ k ≤ 30, -45 ≤ l ≤ 33
Reflections collected	25507
Independent reflections	11515 [R_{int} = 0.0567, R_{sigma} = 0.0708]
Data/restraints/parameters	11515/0/490
Goodness-of-fit on F^2	1.029
Final R indexes [I>=2σ (I)]	R_1 = 0.0593, wR_2 = 0.1311
Final R indexes [all data]	R_1 = 0.0947, wR_2 = 0.1538
Largest diff. peak/hole / e Å^{-3}	1.94/-2.07

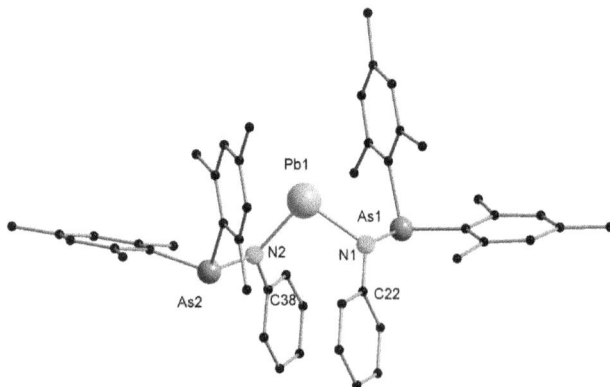

5.2.20 [(Mes$_2$AsNPh)$_2$Ge{Ni(COD)}] (21)

Compound	21
Empirical formula	C$_{56}$H$_{66}$As$_2$GeN$_2$Ni
Formula weight	1048.24
Temperature/K	100
Crystal system	triclinic
Space group	P-1
a/Å	11.534(2)
b/Å	16.018(3)
c/Å	18.389(3)
α/°	102.391(12)
β/°	104.178(11)
γ/°	109.839(12)
Volume/Å^3	2928.7(8)
Z	2
ρ$_{calc}$g/cm^3	1.189
μ/mm^{-1}	1.988
F(000)	1084.0
Crystal size/mm^3	0.26 × 0.097 × 0.016
Radiation	MoKα (λ = 0.71073)
2Θ range for data collection/°	4.746 to 62.088
Index ranges	-15 ≤ h ≤ 16, -23 ≤ k ≤ 20, -26 ≤ l ≤ 26
Reflections collected	27404
Independent reflections	14699 [R$_{int}$ = 0.1450, R$_{sigma}$ = 0.2462]
Data/restraints/parameters	14699/0/571
Goodness-of-fit on F^2	1.028
Final R indexes [I>=2σ (I)]	R$_1$ = 0.1038, wR$_2$ = 0.2183
Final R indexes [all data]	R$_1$ = 0.2144, wR$_2$ = 0.2786
Largest diff. peak/hole / e Å^{-3}	1.01/-1.14

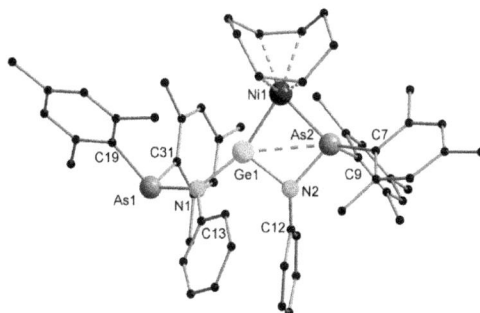

6 Summary

In this thesis, arsinoamide ligands and their metal complexes are reported. First, the aminoarsane Mes$_2$AsN(H)Ph (**1**) was synthesized from Mes$_2$AsCl and aniline in the presence of triethylamine (Scheme 6.1). Treatment of Mes$_2$AsN(H)Ph with ^{n}BuLi, NaN(SiMe$_3$)$_2$ and KH resulted in the first alkali metal complexes of arsinoamides, [(Mes$_2$AsNPh){Li(OEt$_2$)$_2$}] (**2**), [(Mes$_2$AsNPh){Na(OEt$_2$)}]$_2$ (**3**) and [(Mes$_2$AsNPh){K(THF)}]$_2$ (**4**), respectively (Scheme 6.1). X-ray diffraction analyses indicate that the lithium compound **2** forms a monomer, while the sodium and potassium analogues (**3** and **4**) feature dimeric arrangements. All of the arsinoamides in compounds **2**, **3** and **4** adopt a *trans* conformation. In addition, no M-As interaction is observed in these compounds.

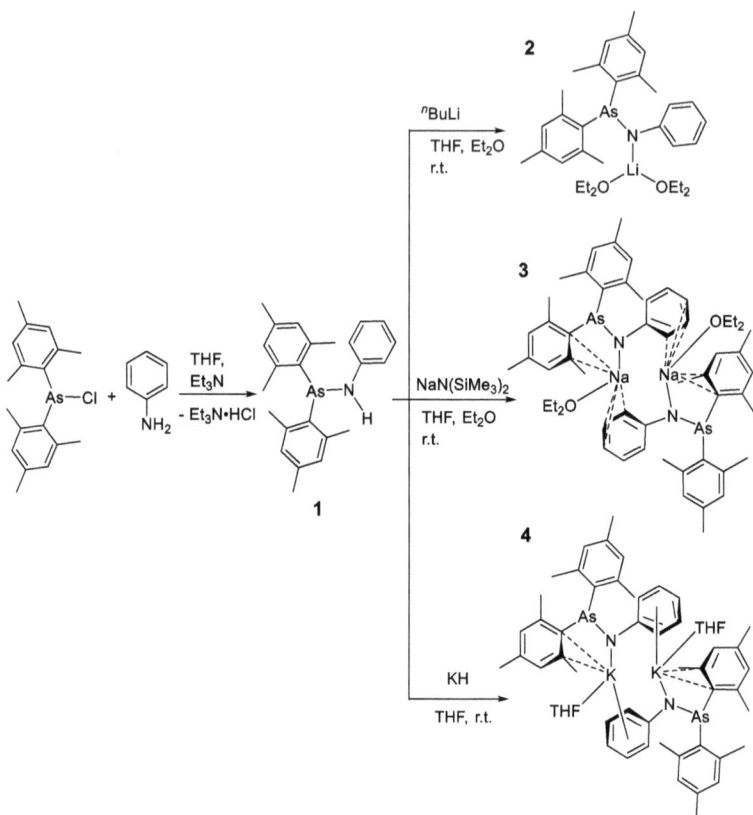

Scheme 6.1 Synthesis of compounds **1-4**.

The second part of this thesis deals with zirconium and hafnium complexes of arsinoamides. Transmetallation of 2 equiv. of compound **2** with appropriate metal halides, *via* salt metathesis, resulted in the corresponding bis-substituted complexes [(Mes$_2$AsNPh)$_2$MCl$_2$(THF)]·(DCM) (**6**: M = Zr, **7**: M = Hf) and [(Mes$_2$AsNPh)$_2$M(NMe$_2$)$_2$] (**8**: M = Zr, **9**: M = Hf) (Scheme 6.2). Owing to the similar ion radii of Zr and Hf, compounds **6** and **7** are isostructural, as well as **8** and **9**. In addition, the M···As interactions in compounds **8** and **9** (less than 3.1 Å) are stronger than those in **6** and **7** (greater than 3.2 Å), which may be attributed to the change of coordination environment at the metal atom. All of the arsinoamides in compounds **6-9** exhibit a *cis* conformation.

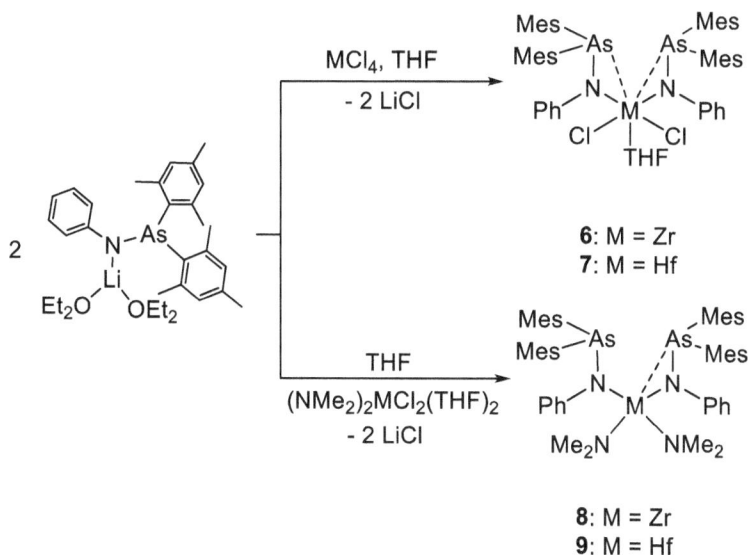

Scheme 6.2 Synthesis of compounds **6-9**.

Third, aluminum and indium complexes of arsinoamides were synthesized and structurally characterized (Scheme 6.3). Using AlCl₃, the mono-substituted complex **12** was obtained *via* salt metathesis as expected. In the case of InCl₃, the ion pair [(Mes₂AsNPh)InCl₃][Li(THF)₄] (**13**) was isolated. The arsinoamides in compounds **12** and **13** adopt a *trans* conformation. X-ray diffraction analyses show that no M-As interaction is observed in compounds **12** and **13**.

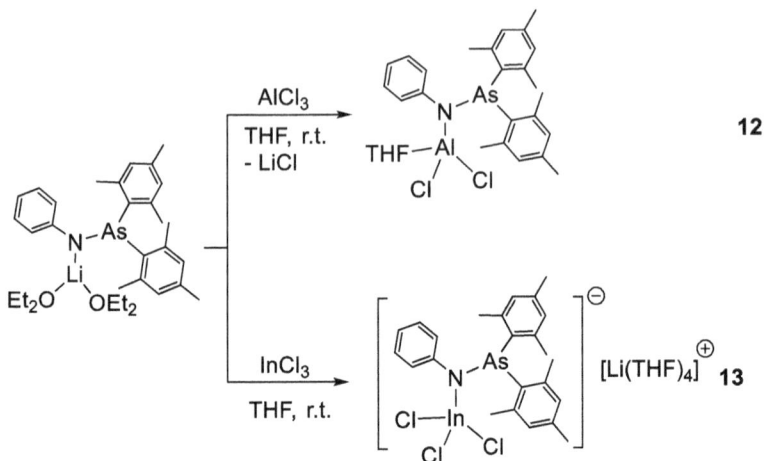

Scheme 6.3 Synthesis of compounds 12 and 13.

Fourth, the synthesis of low-valent group 14 complexes of arsinoamides was investigated, revealing clear trends in reactivity. The benzamidinato-supported silylene ([{PhC(^{t}BuN)$_2$}SiCl]) could insert into the As-N bond under mild conditions, after an oxidative addition step, giving 14 with the first N=Si-As fragment. Owing to a larger singlet-triplet gap, the heavier analogue [{PhC(^{t}BuN)$_2$}GeCl] could not activate the As-N bond, giving complex 15 (Scheme 6.4).

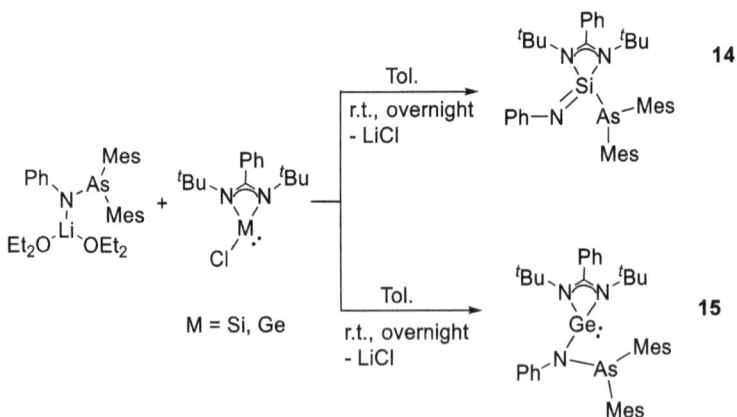

Scheme 6.4 Synthesis of compounds 14 and 15.

To further investigate the As-N bond activation by tetrylenes, transmetallation reactions of lithium arsinoamides with ECl_2 (E = Ge, Sn, Pb) were considered. With 1 equiv. of compound **2**, three different kinds of complexes were isolated and structurally characterized (Scheme 6.5). For $GeCl_2$·dioxane, the germanium atom could insert into the As-N bond, leading to compound **16** ([{Mes_2As}$ClGe(\mu$-NPh)]$_2$) with a Ge_2N_2 core and rare Ge(IV)-As(III) bonds. Compared with [{$PhC(N^tBu)_2$}GeCl], it seems that the substituents on the germylene atom play a vital role in the As-N bond activation. Using $SnCl_2$, the mono-substituted compound **17** ([(Mes_2AsNPh)SnCl(THF)]) was isolated. In the case of $PbCl_2$, only the bis-substituted compound **20** ([(Mes_2AsNPh)$_2$Pb]) was obtained due to the larger ionic radius of the lead atom.

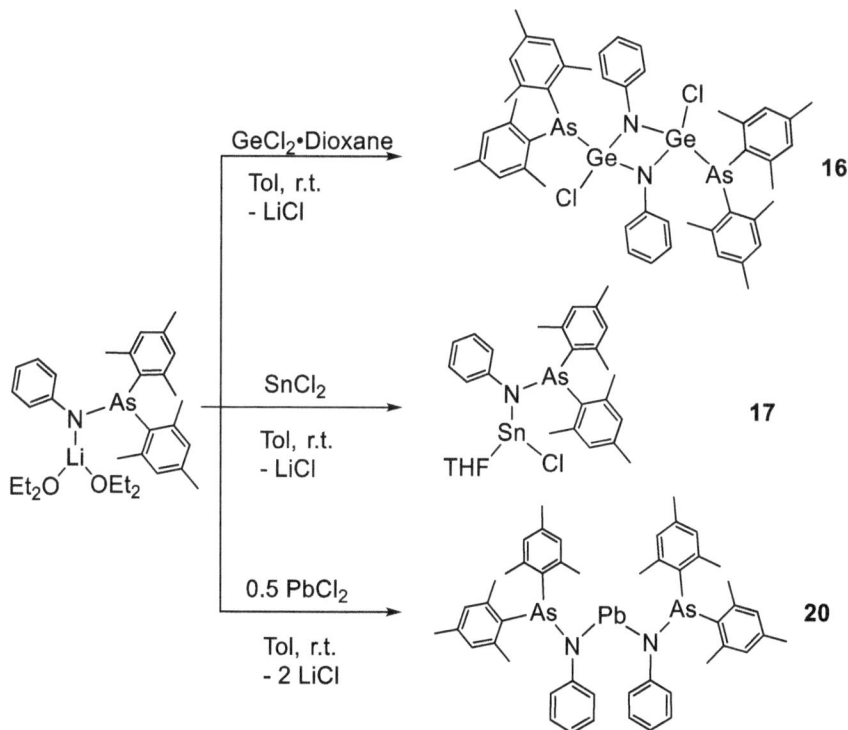

Scheme 6.5 Synthesis of compounds **16**, **17** and **20**.

Using 2 equiv. of compound **2**, the three homoleptic bis-substituted complexes [(Mes_2AsNPh)$_2$E] (E = Ge(**18**), Sn(**19**) and Pb(**20**)) were isolated and structurally characterized (Scheme 6.6).

According to the X-ray diffraction analysis, complexes **18**, **19** and **20** are isostructural. No As-N bond activation or E-As interaction is observed in these compounds.

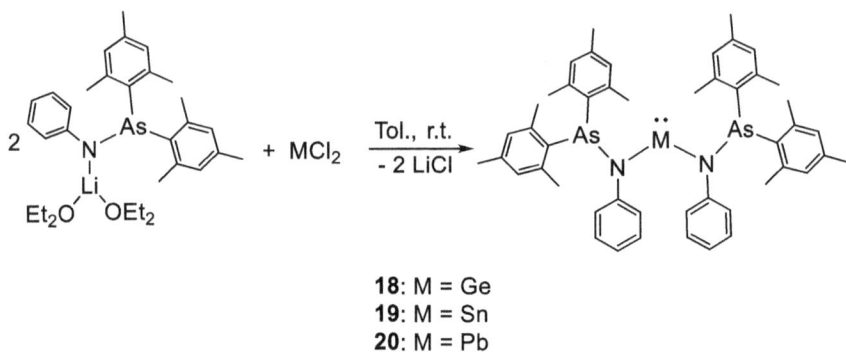

18: M = Ge
19: M = Sn
20: M = Pb

Scheme 6.6 Synthesis of compounds **18**, **19** and **20**.

The first arsinoamide-supported germylene-nickel complex **21** was synthesized through the reaction of compound **18** with Ni(COD)₂ (Scheme 6.7). After the leaving of one COD molecule, the nickel center coordinates to one germanium and one arsenic atom. As result, a sharp decrease in one of the As-N-Ge bond angles can be observed. In addition, a weak Ge⋯As interaction and a conformational interconversion of one arsinoamide ligand could be detected in **21**.

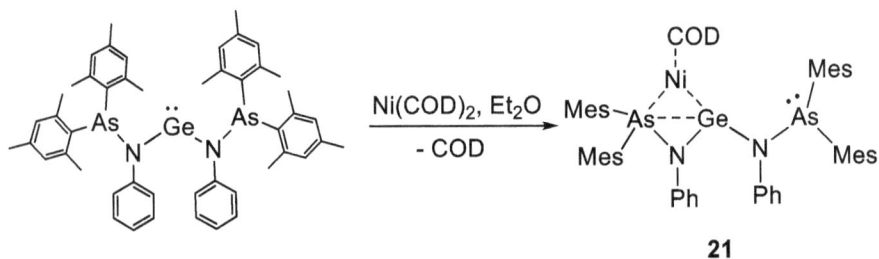

Scheme 6.7 Synthesis of compound **21**.

6 Zusammenfassung

In dieser Arbeit wird die Synthese von Arsinoamiden und ihren Metallkomplexen beschrieben. Im ersten Teil dieser Arbeit wurde das Aminoarsan $Mes_2AsN(H)Ph$ (**1**) aus Mes_2AsCl und Anilin in Gegenwart von Triethylamin synthetisiert (Schema 6.1). Durch die Umsetzung von $Mes_2AsN(H)Ph$ mit ^nBuLi, $NaN(SiMe_3)_2$ und KH konnten die ersten Alkalimetalkomplexen von Arsinoamiden $[(Mes_2AsNPh)\{Li(OEt_2)_2\}]$ (**2**), $[(Mes_2AsNPh)\{Na(OEt_2)\}]_2$ (**3**), $[(Mes_2AsNPh)\{K(THF)\}]_2$ (**4**) erhalten werden (Schema 6.1). Die Ergebnisse der Röntgenstrukturanalyse zeigten, dass die Lithiumverbindung als Monomer vorliegt, während die Natrium- und Kaliumverbindungen dimere Strukturen ausbilden. Alle Arsinoamide in **2**, **3** und **4** liegen in *trans*-Konformation vor. In keiner der gezeigten Kristallstrukturen wurde eine Metall-Arsen-Wechselwirkung beobachtet.

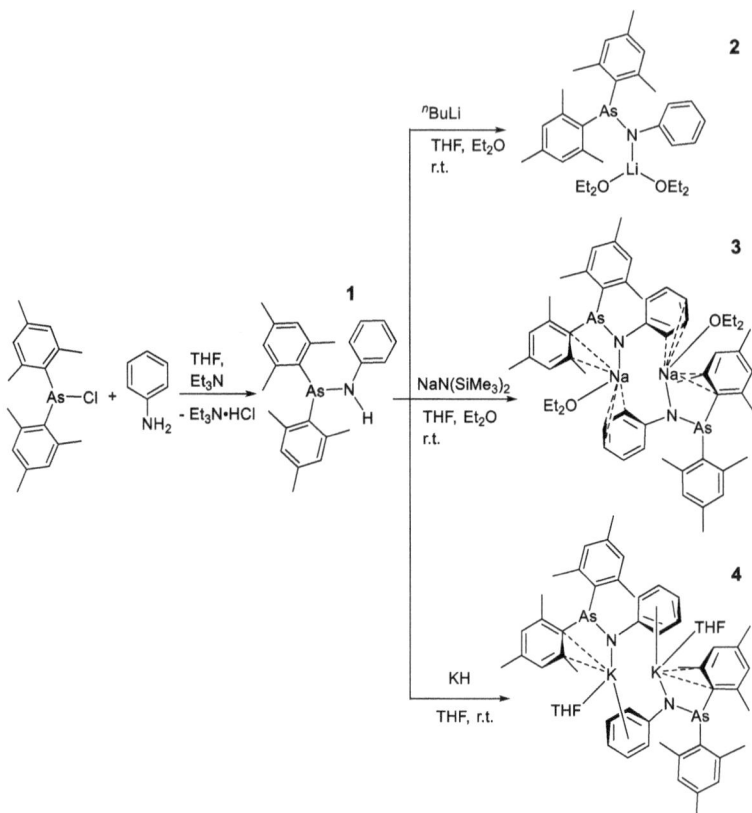

Schema 6.1 Synthese der Verbindungen **1-4**.

Im zweiten Teil dieser Arbeit wurden Zirkonium- und Hafniumkomplexe von Arsinoamiden dargestellt und untersucht. Die Umsetzung der entsprechenden Metallvorläufer mit 2 Äquiv. von **2** führte zu den zweifachsubstituierten Komplexen [(Mes$_2$AsNPh)$_2$MCl$_2$(THF)]·(DCM) (**6**: M = Zr; **7**: M = Hf) und [(Mes$_2$AsNPh)$_2$M(NMe$_2$)$_2$] (**8**: M = Zr; **9**: M = Hf) (Schema 6.2). Die Röntgenstrukturanalysen haben ergeben, dass Verbindung **6** isostrukturell zu **7** und **8** isostrukturell zu **9** ist. Die M-As-Wechselwirkungen in Verbindungen **8** und **9** (kleiner als 3.1 Å) sind stärker als die in Verbindungen **6** und **7** (größer als 3.2 Å). Der kürzere M-As-Abstand kann zur Änderung der Elektronendichte und der Koordinationsumgebung am Metallatom beitragen. Die Arsinoamiden in **6-9** liegen in *trans*-Konformation vor.

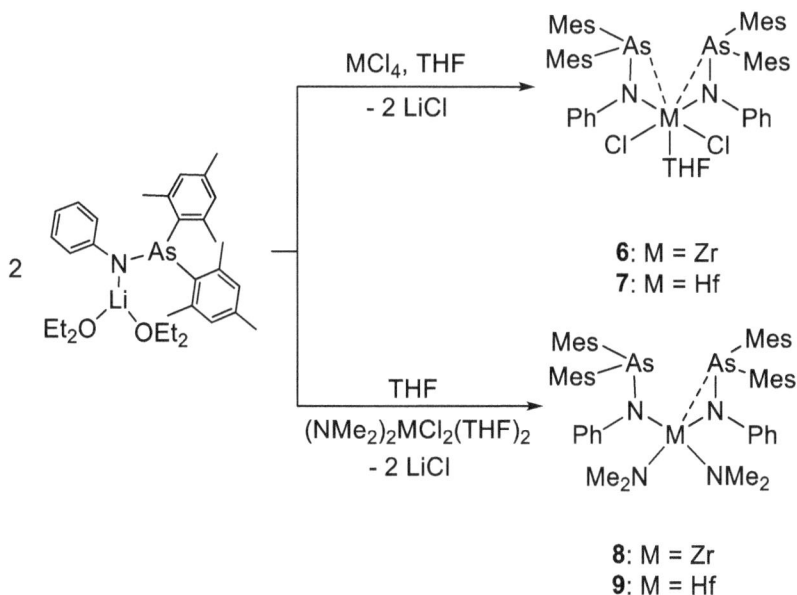

Schema 6.2 Synthese der Verbindungen **6-9**.

Des Weiteren wurden Aluminium- und Indiumkomplexe mit dem Arsinoamid synthetisiert (Schema 6.3). Mit AlCl$_3$ wurde der erwartete, einfach substituierte Komplex **12** erhalten. Die ionische Verbindung [(Mes$_2$AsNPh)InCl$_3$][Li(THF)$_4$] (**13**) wurde mit InCl$_3$ synthetisiert. In den Verbindung **12** und **13** liegen keine M-As Wechselwirkungen vor. Die Arsinoamiden in **12** und **13** zeigen eine *trans*-Konformation.

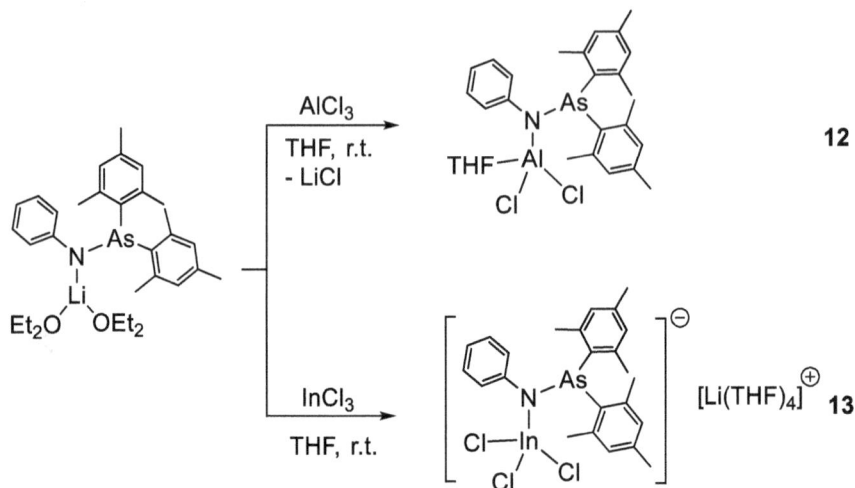

Schema 6.3 Synthese der Verbindungen **12** und **13**.

Im vierten Teil wurden niedervalente Verbindungen der Elemente der Gruppe 14 mit Arsinoamiden untersucht. Das durch einen Benzamidinato-Liganden stabilisierte Silylen ([{PhC(^{t}BuN)$_2$}SiCl]) konnte unter milden Reaktionsbedingungen in die As-N Bindung eingeführt werden, wodurch Komplex **14** und das erste N=Si-As Fragment erhalten wurde. Aufgrund der Zunahme der Singulett-Triplett-Lücke konnte das schwerere Analog [{PhC(^{t}BuN)$_2$}GeCl] die As-N Bindung nicht aktivieren (Schema 6.4), wodurch Komplex **15** erhalten wurde.

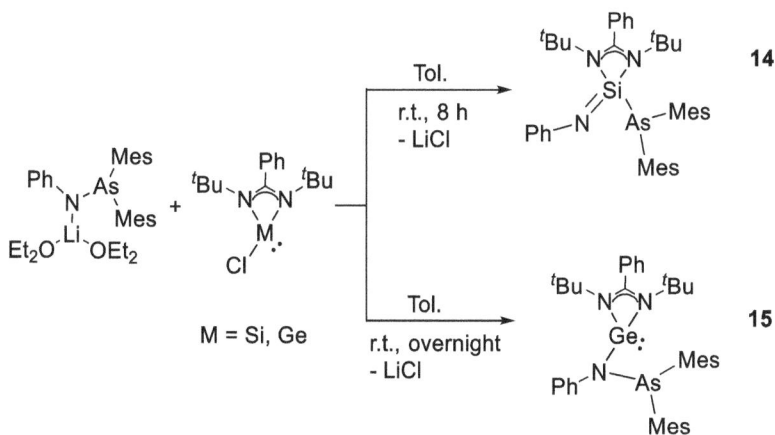

Schema 6.4 Synthese der Verbindungen **14** und **15**.

Zur weiteren Untersuchung der As-N-Bindungsaktivierung wurden mit den Dichloride der 14. Gruppe ECl$_2$ (E = Ge, Sn, Pb) mit dem Arsinoamid umgesetzt. Durch die Umsetzung von 1 Äquiv. von **2** wurden drei unterschiedliche Arten von Komplexen isoliert. GeCl$_2$·Dioxan konnte in die As-N-Bindung eingeführt werden. Als Produkte wurden Komplex **16** ([{Mes$_2$As}ClGe(μ-NPh)]$_2$) mit einem Ge$_2$N$_2$ Ring und einer seltenen Ge(IV)-As(III) Bindung erhalten. Im Vergleich zum [{PhC(N^{t}Bu)$_2$}GeCl], spielt der substituierte Ligand im Germylen in der As-N-Bindungsaktivierung eine entscheidende Rolle. Mit SnCl$_2$ wurde wie erwartet, der einfachsubstituierter Komplex **17** [(Mes$_2$AsNPh)SnCl(THF)] erhalten. Mit PbCl$_2$ konnte aufgrund der Radiuszunahme von Blei nur die zweifach substituierte Komplex **20** [(Mes$_2$AsNPh)$_2$Pb] isoliert werden (Schema 6.5).

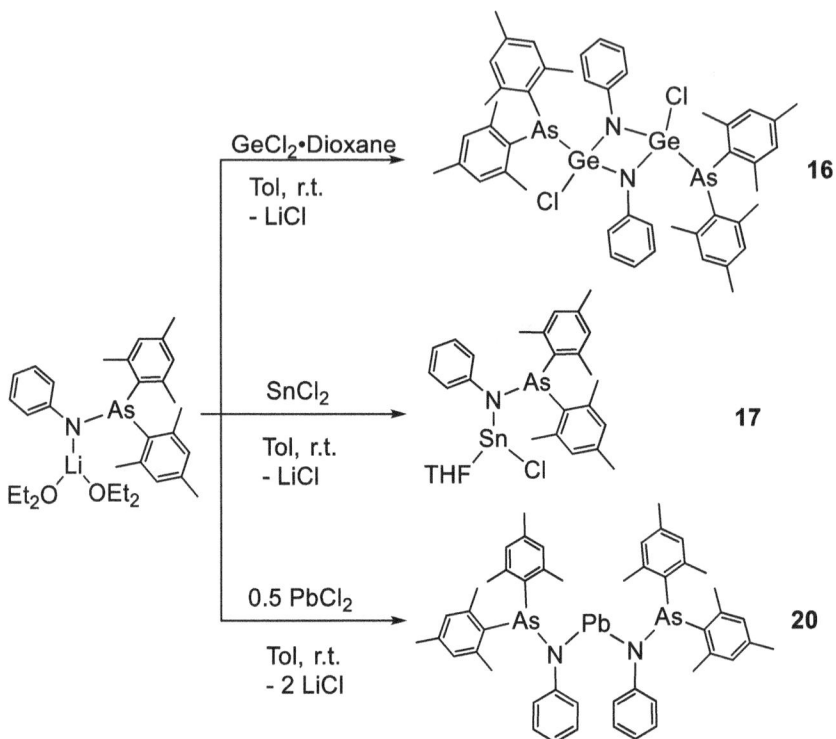

Schema 6.5 Synthese der Verbindungen **16**, **17** und **20**.

Mit 2 Äquiv. von **2** konnten zweifach substituierte Komplexe [(Mes₂AsNPh)₂E] (E = Ge (**18**), Sn (**19**) and Pb (**20**)) synthetisiert werden. Die Molekülstrukturen im Festkörper zeigen das die Verbindungen isostrukturell sind. Zudem wurden keine As-N-Bindungsaktivierung beobachtet.

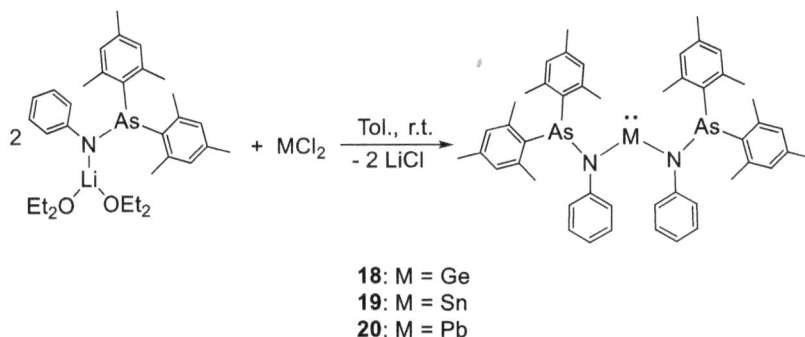

18: M = Ge
19: M = Sn
20: M = Pb

Schema 6.6 Synthese der Verbindungen **18-20**.

Die erste Arsinoamid stabilisierte Germylen-Nickel Verbindung **21** wurde durch die Reaktion von **18** mit Ni(COD)₂ erhalten (Schema 6.7). Nach der Abspaltung von einem COD-Molekül, koordiniert Nickel sowohl an das Germaniumatom- als auch an das Arsenatom. Im Vergleich zu **18** konnte eine starke Abnahme des As-N-Ge-Bindungswinkel beobachtet werden. Zusätzlich gibt es eine schwache Ge···As Wechselwirkungen und eine Konformationsumwandlung des Arsinoamids in **21**.

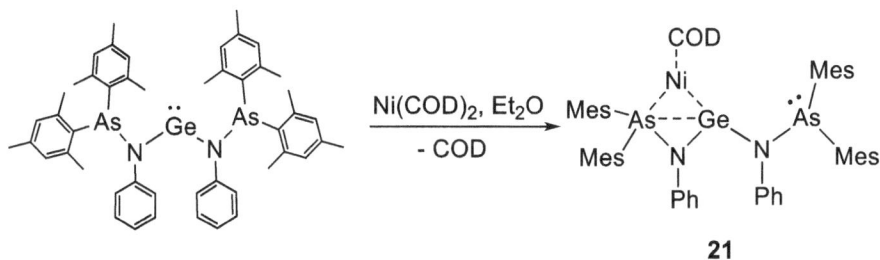

21

Scheme 6.1 Synthese der Verbindung **21**.

7 References

[1]   A. Michaelis, R. Kaehne, *Ber. Dtsch. Chem. Ges.* **1898**, *31*, 1048.

[2]   A. Michaelis, *Justus Liebigs Annalen der Chemie* **1903**, *326*, 129.

[3]   G. Ewart, D. Payne, A. Porte, A. Lane, *J. Chem. Soc.* **1962**, 3984.

[4]   A. B. Burg, P. J. Slota Jr, *J. Am. Chem. Soc.* **1958**, *80*, 1107.

[5]   H. H. Sisler, N. L. Smith, *J.Org.Chem* **1961**, *26*, 4733.

[6]   G. Ewart, A. Lane, *J. Chem. Soc* **1964**, *1543*.

[7]   D. Fenske, B. Maczek, K. Maczek, *Z. Anorg. Allg. Chem.* **1997**, *623*, 1113.

[8]   H. Nöth, L. Meinel, *Z. Anorg. Allg. Chem.* **1967**, *349*, 225.

[9]   C. Fliedel, A. Ghisolfi, P. Braunstein, *Chem. Rev.* **2016**, *116*, 9237.

[10]   H. Schmidbaur, S. Lauteschläger, F. H. Köhler, *J. Organomet. Chem.* **1984**, *271*, 173.

[11]   M. L. Shozi, H. B. Friedrich, *S. Afr. J. Chem.* **2012**, *65*, 214.

[12]   S. A. Katz, V. S. Allured, A. D. Norman, *Inorg. Chem.* **1994**, *33*, 1762.

[13]   I. Abd-Ellah, E. Ibrahim, A. El-khazander, *Phosphorus, Sulfur Silicon Relat. Elem.* **1987**, *31*, 13.

[14]   R. A. Shaw, *Phosphorus, Sulfur Silicon Relat. Elem.* **1978**, *4*, 101.

[15]   G. Trinquier, M. T. Ashby, *Inorg. Chem.* **1994**, *33*, 1306.

[16]   N. Burford, T. S. Cameron, K. D. Conroy, B. Ellis, M. Lumsden, C. L. Macdonald, R. McDonald, A. D. Phillips, P. J. Ragogna, R. W. Schurko, *J. Am. Chem. Soc.* **2002**, *124*, 14012.

[17]   M. Witt, H. W. Roesky, *Chem. Rev.* **1994**, *94*, 1163.

[18]   J. M. Barendt, E. G. Bent, R. C. Haltiwanger, A. D. Norman, *Inorg. Chem.* **1989**, *28*, 2334.

[19]   S. Krishnamurthy, *Phosphorus, Sulfur, Silicon Relat. Elem.* **1994**, *87*, 101.

[20]   K. Izod, *Adv. Inorg. Chem.* **2000**, *50*, 33.

[21]   A. H. Cowley, M. J. Dewar, W. R. Jackson Jr, W. B. Jennings, *J. Am. Chem. Soc.* **1970**, *92*, 5206.

[22]   M. T. Ashby, Z. Li, *Inorg. Chem.* **1992**, *31*, 1321.

[23]   A. E. Reed, P. v. R. Schleyer, *J. Am. Chem. Soc.* **1990**, *112*, 1434.

[24]   I. V. Alabugin, M. Manoharan, S. Peabody, F. Weinhold, *J. Am. Chem. Soc.* **2003**, *125*, 5973.

[25]   N. Poetschke, M. Nieger, M. A. Khan, E. Niecke, M. T. Ashby, *Inorg. Chem.* **1997**, *36*, 4087.

[26]   P. W. Roesky, *Heteroat. Chem.* **2002**, *13*, 514.

[27]   Z. Fei, P. J. Dyson, *Coord. Chem. Rev.* **2005**, *249*, 2056.

[28]   M. Balakrishna, V. S. Reddy, S. Krishnamurthy, J. Nixon, J. B. S. Laurent, *Coord. Chem. Rev.* **1994**, *129*, 1.

[29]   T. Appleby, J. D. Woollins, *Coord. Chem. Rev.* **2002**, *235*, 121.

[30]   Y.-T. Bi, L. Li, Y.-R. Guo, Q.-J. Pan, *Inorg. Chem.* **2019**, *2*, 1290.

[31]   P. W. Smith, S. R. Ellis, R. C. Handford, T. D. Tilley, *Organometallics* **2019**, *2*, 336.

[32]   A. J. Ayres, M. Zegke, J. P. Ostrowski, F. Tuna, E. J. McInnes, A. J. Wooles, S. T. Liddle, *Chem. Commun.* **2018**, *54*, 13515.

[33]   G. Feng, M. Zhang, D. Shao, X. Wang, S. Wang, L. Maron, C. Zhu, *Nat. Chem.* **2019**, 1.

[34]   O. Kühl, T. Koch, F. B. Somoza, P. C. Junk, E. Hey-Hawkins, D. Plat, M. S. Eisen, *J. Organomet. Chem.* **2000**, *604*, 116.

[35]   V. V. Kotov, E. V. Avtomonov, J. Sundermeyer, K. Harms, D. A. Lemenovskii, *Eur. J. Inorg. Chem.* **2002**, *2002*, 678.

[36]   P. W. Roesky, M. T. Gamer, M. Puchner, A. Greiner, *Chem. -Eur. J.* **2002**, *8*, 5265.

[37]   C. M. Thomas, *Comments Inorg. Chem.* **2011**, *32*, 14.

[38]   M. R. Crawley, A. E. Friedman, T. R. Cook, *Inorg. Chem.* **2018**, *57*, 5692.

[39]   L. Fohlmeister, A. Stasch, *Chem. -Eur. J.* **2016**, *22*, 10235.

[40]   Y. Chen, D. Song, J. Li, X. Hu, X. Bi, T. Jiang, Z. Hou, *ChemCatChem* **2018**, *10*, 159.

[41]   P. A. Aguirre, C. A. Lagos, S. A. Moya, C. Zúñiga, C. Vera-Oyarce, E. Sola, G. Peris, J. C. Bayón, *Dalton Trans.* **2007**, 5419.

[42]   D. Benito-Garagorri, K. Kirchner, *Acc. Chem. Res.* **2008**, *41*, 201.

[43]   F. Völcker, Y. Lan, A. K. Powell, P. W. Roesky, *Dalton Trans.* **2013**, *42*, 11471.

[44]   S. Kuppuswamy, M. W. Bezpalko, T. M. Powers, M. M. Turnbull, B. M. Foxman, C. M. Thomas, *Inorg. Chem.* **2012**, *51*, 8225.

[45]   N. I. Saper, M. W. Bezpalko, B. M. Foxman, C. M. Thomas, *Polyhedron* **2016**, *114*, 88.

[46]   S. C. Tobias, R. F. Borch, *J. Med. Chem.* **2001**, *44*, 4475.

[47]   M. Işıklan, N. Asmafiliz, E. E. Özalp, E. E. Ilter, Z. Kılıç, B. n. Çoşut, S. Yeşilot, A. Kılıç, A. Öztürk, T. Hökelek, *Inorg. Chem.* **2010**, *49*, 7057.

[48]   G. Mutlu, G. Elmas, Z. Kılıç, T. Hökelek, L. Y. Koc, M. Türk, L. Açık, B. Aydın, H. Dal, *Inorg. Chim. Acta* **2015**, *436*, 69.

[49]   T. G. Wetzel, S. Dehnen, P. W. Roesky, *Angew. Chem. Int. Ed.* **1999**, *38*, 1086.

[50]   F. Völcker, F. M. Mück, K. D. Vogiatzis, K. Fink, P. W. Roesky, *Chem. Commun.* **2015**, *51*, 11761.

[51]   F. Völcker, P. W. Roesky, *Dalton Trans.* **2016**, *45*, 9429.

[52]   A. M. Baranger, F. J. Hollander, R. G. Bergman, *J. Am. Chem. Soc.* **1993**, *115*, 7890.

[53]   H. Tsutsumi, Y. Sunada, Y. Shiota, K. Yoshizawa, H. Nagashima, *Organometallics* **2009**, *28*, 1988.

[54]   B. P. Greenwood, G. T. Rowe, C.-H. Chen, B. M. Foxman, C. M. Thomas, *J. Am. Chem. Soc.* **2009**, *132*, 44.

[55]   Y. Sunada, T. Sue, T. Matsumoto, H. Nagashima, *J. Organomet. Chem.* **2006**, *691*, 3176.

[56]   J. W. Napoline, S. J. Kraft, E. M. Matson, P. E. Fanwick, S. C. Bart, C. M. Thomas, *Inorg. Chem.* **2013**, *52*, 12170.

[57]   J. P. Krogman, J. R. Gallagher, G. Zhang, A. S. Hock, J. T. Miller, C. M. Thomas, *Dalton Trans.* **2014**, *43*, 13852.

[58]   N. R. Halcovitch, M. D. Fryzuk, *Organometallics* **2013**, *32*, 5705.

[59]   C. Qi, S. Zhang, J. Sun, *Appl. Organomet. Chem.* **2006**, *20*, 138.

[60]   B. Wu, R. I. Hernández Sánchez, M. W. Bezpalko, B. M. Foxman, C. M. Thomas, *Inorg. Chem.* **2014**, *53*, 10021.

[61]   S. Kuppuswamy, B. G. Cooper, M. W. Bezpalko, B. M. Foxman, T. M. Powers, C. M. Thomas, *Inorg. Chem.* **2012**, *51*, 1866.

[62]   B. Wu, M. W. Bezpalko, B. M. Foxman, C. M. Thomas, *Chem. Sci.* **2015**, *6*, 2044.

[63]   W. K. Walker, D. L. Anderson, R. W. Stokes, S. J. Smith, D. J. Michaelis, *Org. Lett.* **2015**, *17*, 752.

[64]   H. Nagashima, T. Sue, T. Oda, A. Kanemitsu, T. Matsumoto, Y. Motoyama, Y. Sunada, *Organometallics* **2006**, *25*, 1987.

[65]   M. Wiecko, D. Girnt, M. Rastätter, T. K. Panda, P. W. Roesky, *Dalton Trans.* **2005**, 2147.

[66]   T. Sue, Y. Sunada, H. Nagashima, *Eur. J. Inorg. Chem.* **2007**, *2007*, 2897.

[67]   W. Zhou, N. I. Saper, J. P. Krogman, B. M. Foxman, C. M. Thomas, *Dalton Trans.* **2014**, *43*, 1984.

[68]   B. G. Cooper, C. M. Fafard, B. M. Foxman, C. M. Thomas, *Organometallics* **2010**, *29*, 5179.

[69]   B. P. Greenwood, S. I. Forman, G. T. Rowe, C.-H. Chen, B. M. Foxman, C. M. Thomas, *Inorg. Chem.* **2009**, *48*, 6251.

[70]   H. Zhang, B. Wu, S. L. Marquard, E. D. Litle, D. A. Dickie, M. W. Bezpalko, B. M. Foxman, C. M. Thomas, *Organometallics* **2017**, *36*, 3498.

[71]   H. S. Zijlstra, J. Pahl, J. Penafiel, S. Harder, *Dalton Trans.* **2017**, *46*, 3601.

[72]   K. Jaiswal, B. Prashanth, D. Bawari, S. Singh, *Eur. J. Inorg. Chem.* **2015**, *2015*, 2565.

[73]   C. W. So, H. W. Roesky, J. Magull, R. B. Oswald, *Angew. Chem. Int. Ed.* **2006**, *45*, 3948.

[74] S. S. Sen, H. W. Roesky, D. Stern, J. Henn, D. Stalke, *J. Am. Chem. Soc.* **2009**, *132*, 1123.

[75] S. Nagendran, S. S. Sen, H. W. Roesky, D. Koley, H. Grubmüller, A. Pal, R. Herbst-Irmer, *Organometallics* **2008**, *27*, 5459.

[76] S. S. Sen, M. P. Kritzler - Kosch, S. Nagendran, H. W. Roesky, T. Beck, A. Pal, R. Herbst - Irmer, *Eur. J. Inorg. Chem.* **2010**, *2010*, 5304.

[77] A. Jana, P. P. Samuel, G. p. Tavčar, H. W. Roesky, C. Schulzke, *J. Am. Chem. Soc.* **2010**, *132*, 10164.

[78] M. Asay, C. Jones, M. Driess, *Chem. Rev.* **2010**, *111*, 354.

[79] A. Jana, C. Schulzke, H. W. Roesky, *J. Am. Chem. Soc.* **2009**, *131*, 4600.

[80] S. K. Mandal, H. W. Roesky, *Acc. Chem. Res.* **2011**, *45*, 298.

[81] J. Schneider, K. M. Krebs, S. Freitag, K. Eichele, H. Schubert, L. Wesemann, *Chem. -Eur. J.* **2016**, *22*, 9812.

[82] A. Schnepf, *Chem. Soc. Rev.* **2007**, *36*, 745.

[83] J. A. Clyburne, N. McMullen, *Coord. Chem. Rev.* **2000**, *210*, 73.

[84] F. Breher, *Coord. Chem. Rev.* **2007**, *251*, 1007.

[85] D. A. Atwood, *Coord. Chem. Rev.* **1997**, *165*, 267.

[86] J. Casas, M. Garcıa-Tasende, J. Sordo, *Coord. Chem. Rev.* **2000**, *209*, 197.

[87] P. Jutzi, N. Burford, *Chem. Rev.* **1999**, *99*, 969.

[88] C. J. Levy, R. J. Puddephatt, *J. Am. Chem. Soc.* **1997**, *119*, 10127.

[89] A. V. Protchenko, J. I. Bates, L. M. Saleh, M. P. Blake, A. D. Schwarz, E. L. Kolychev, A. L. Thompson, C. Jones, P. Mountford, S. Aldridge, *J. Am. Chem. Soc.* **2016**, *138*, 4555.

[90] P. B. Hitchcock, H. A. Jasim, M. F. Lappert, W.-P. Leung, A. K. Rai, R. E. Taylor, *Polyhedron* **1991**, *10*, 1203.

[91] T. Chu, G. I. Nikonov, *Chem. Rev.* **2018**, *118*, 3608.

[92] S. Grélaud, P. Cooper, L. J. Feron, J. F. Bower, *J. Am. Chem. Soc.* **2018**, *140*, 9351.

[93] J. Wu, T.-L. Yu, C.-T. Chen, C.-C. Lin, *Coord. Chem. Rev.* **2006**, *250*, 602.

[94] T. J. Hadlington, M. Hermann, G. Frenking, C. Jones, *J. Am. Chem. Soc.* **2014**, *136*, 3028.

[95] Y. Wu, C. Shan, Y. Sun, P. Chen, J. Ying, J. Zhu, L. L. Liu, Y. Zhao, *Chem. Commun.* **2016**, *52*, 13799.

[96] W. A. Herrmann, R. W. Fischer, J. D. Correia, *J. Mol. Catal.* **1994**, *94*, 213.

[97] T. J. Hadlington, M. Driess, C. Jones, *Chem. Soc. Rev.* **2018**, *47*, 4176.

[98] G. H. Spikes, J. C. Fettinger, P. P. Power, *J. Am. Chem. Soc.* **2005**, *127*, 12232.

[99] T. J. Hadlington, in *On the Catalytic Efficacy of Low-Oxidation State Group 14 Complexes*, Springer, **2017**, pp. 147.

[100] M. K. Sharma, D. Singh, P. Mahawar, R. Yadav, S. Nagendran, *Dalton Trans.* **2018**, *47*, 5943.

[101] N. Parvin, S. Pal, V. C. Rojisha, S. De, P. Parameswaran, S. Khan, *ChemistrySelect* **2016**, *1*, 1991.

[102] S. Khan, S. Pal, N. Kathewad, I. Purushothaman, S. De, P. Parameswaran, *Chem. Commun.* **2016**, *52*, 3880.

[103] C.-W. So, M. L. B. Ismail, *Chem. Commun.* **2019**, *55*, 2074.

[104] T. Böttcher, C. Jones, *Main Group Met. Chem.* **2015**, *38*, 165.

[105] S. Pal, R. Dasgupta, S. Khan, *Organometallics* **2016**, *35*, 3635.

[106] D. Olbert, A. Kalisch, H. Görls, I. M. Ondik, M. Reiher, M. Westerhausen, *Z. Anorg. Allg. Chem.* **2009**, *635*, 462.

[107] W. Cullen, H. Eméléus, *J. Chem. Soc.* **1959**, 372.

[108] F. Yambushev, N. K. Tenisheva, K. Z. Khusainov, *Chem. Infor.* **1977**, *8*, 12.

[109] R. A. Zingaro, K. J. Irgolic, *J. Organomet. Chem.* **1977**, *138*, 125.

[110] F. Kober, *Synthesis* **1982**, *1982*, 173.

[111] F. Reiß, A. Schulz, A. Villinger, N. Weding, *Dalton Trans.* **2010**, *39*, 9962.

[112] G. Kokorev, F. Yambushev, L. Al'Metkina, *Chem. Infor.* **1988**, *19*, no.

[113] J. Wardell, *Organomet. Chem.* **1984**, *12*, 127.

[114] W. Kutzelnigg, *Angew. Chem. Int. Ed.* **1984**, *23*, 272.

[115]   M. Dub, *Compounds of Arsenic, Antimony, and Bismuth, Vol. 3*, Springer Science & Business Media, **2013**.

[116]   N. C. Norman, *Chemistry of arsenic, antimony and bismuth*, Springer Science & Business Media, **1997**.

[117]   W. Cullen, J. Trotter, *Can. J. Chem.* **1961**, *39*, 2602.

[118]   F. Blicke, F. Smith, *J. Am. Chem. Soc.* **1929**, *51*, 1558.

[119]   T. R. Harper, N. W. Kingham, *Water Environ. Res* **1992**, *64*, 200.

[120]   E. O. Kartinen Jr, C. J. Martin, *Desalination* **1995**, *103*, 79.

[121]   Y.-h. Xu, T. Nakajima, A. Ohki, *J. Hazard. Mater.* **2002**, *92*, 275.

[122]   L. C. Roberts, S. J. Hug, T. Ruettimann, M. M. Billah, A. W. Khan, M. T. Rahman, *Environ. Sci Technol.* **2004**, *38*, 307.

[123]   C. G. Pitt, A. P. Purdy, K. T. Higa, R. L. Wells, *Organometallics* **1986**, *5*, 1266.

[124]   A. Hinz, A. Schulz, A. Villinger, *Chem. -Eur. J.* **2016**, *22*, 12266.

[125]   O. Kühl, S. Blaurock, J. Sieler, E. Hey-Hawkins, *Polyhedron* **2001**, *20*, 111.

[126]   A. Schulz, P. Mayer, A. Villinger, *Inorg. Chem.* **2007**, *46*, 8316.

[127]   M. Driess, H. Pritzkow, S. Rell, U. Winkler, *Organometallics* **1996**, *15*, 1845.

[128]   A. Stasch, *Angew. Chem.* **2012**, *124*, 1966.

[129]   B. Wrackmeyer, C. Schödel, R. Kempe, G. Glatz, A. Noor, *Z. Anorg. Allg. Chem.* **2016**, *642*, 922.

[130]   D. F.-J. Piesik, P. Haack, S. Harder, C. Limberg, *Inorg. Chem.* **2009**, *48*, 11259.

[131]   D. L. Clark, J. C. Gordon, J. C. Huffman, R. L. Vincent-Hollis, J. G. Watkin, B. D. Zwick, *Inorg. Chem.* **1994**, *33*, 5903.

[132]   R. García-Rodríguez, H. R. Simmonds, D. S. Wright, *Organometallics* **2014**, *33*, 7113.

[133]   E. Witt, D. W. Stephan, *Inorg. Chem.* **2001**, *40*, 3824.

[134]   C. D. Carmichael, M. D. Fryzuk, *Dalton Trans.* **2005**, 452.

[135]   X. Chen, M. T. Gamer, P. W. Roesky, *Dalton Trans.* **2018**, *47*, 12521.

[136]   W. Levason, M. L. Matthews, B. Patel, G. Reid, M. Webster, *Dalton Trans.* **2004**, 3305.

[137]   M. Schmidt, A. E. Seitz, M. Eckhardt, G. b. Balázs, E. V. Peresypkina, A. V. Virovets, F. Riedlberger, M. Bodensteiner, E. M. Zolnhofer, K. Meyer, *J. Am. Chem. Soc.* **2017**, *139*, 13981.

[138]   L. Pauling, *The Nature of the Chemical Bond, Vol. 260*, Cornell university press Ithaca, NY, **1960**.

[139]   B. Cordero, V. Gómez, A. E. Platero-Prats, M. Revés, J. Echeverría, E. Cremades, F. Barragán, S. Alvarez, *Dalton Trans.* **2008**, 2832.

[140]   E. Smolensky, M. Kapon, M. S. Eisen, *Organometallics* **2007**, *26*, 4510.

[141]   C. M. Thomas, J. W. Napoline, G. T. Rowe, B. M. Foxman, *Chem. Commun.* **2010**, *46*, 5790.

[142]   Y. Zou, D. Wang, K. Wurst, C. Kühnel, I. Reinhardt, U. Decker, V. Gurram, S. Camadanli, M. R. Buchmeiser, *Chem. -Eur. J.* **2011**, *17*, 13832.

[143]   R. M. Gauvin, J. A. Osborn, J. Kress, *Organometallics* **2000**, *19*, 2944.

[144]   A. Novak, A. J. Blake, C. Wilson, J. B. Love, *Chem. Commun.* **2002**, 2796.

[145]   X. Yu, S.-J. Chen, X. Wang, X.-T. Chen, Z.-L. Xue, *Organometallics* **2009**, *28*, 4269.

[146]   F. Lindenberg, J. Sieler, E. Hey‑Hawkins, U. Müller, A. Pilz, *Z. Anorg. Allg. Chem.* **1996**, *622*, 683.

[147]   M. J. Sgro, D. W. Stephan, *Chem. Commun.* **2013**, *49*, 2610.

[148]   V. N. Setty, W. Zhou, B. M. Foxman, C. M. Thomas, *Inorg. Chem.* **2011**, *50*, 4647.

[149]   S. Brenner, R. Kempe, P. Arndt, *Z. Anorg. Allg. Chem.* **1995**, *621*, 2021.

[150]   L. T. Elrod, H. Boxwala, H. Haq, A. W. Zhao, R. Waterman, *Organometallics* **2012**, *31*, 5204.

[151]   E. Hey-Hawkins, F. Lindenberg, *Organometallics* **1994**, *13*, 4643.

[152]   A. J. Roering, J. J. Davidson, S. N. MacMillan, J. M. Tanski, R. Waterman, *Dalton Trans.* **2008**, 4488.

[153]   N. R. Halcovitch, M. D. Fryzuk, *Aust. J. Chem.* **2016**, *69*, 555.

[154]   K.-C. Yang, C.-C. Chang, J.-Y. Huang, C.-C. Lin, G.-H. Lee, Y. Wang, M. Y. Chiang, *J. Organomet. Chem.* **2002**, *648*, 176.

[155]   A. Stasch, *Angew. Chem.* **2014**, *126*, 10364.

[156]   J. Vrána, R. Jambor, A. Růžička, M. Alonso, F. De Proft, A. Lyčka, L. Dostál, *Dalton Trans.* **2015**, *44*, 4533.

[157]   R. Neufeld, T. L. Teuteberg, R. Herbst-Irmer, R. A. Mata, D. Stalke, *J. Am. Chem. Soc.* **2016**, *138*, 4796.

[158]   S. Yuan, S. Bai, D. Liu, W.-H. Sun, *Organometallics* **2010**, *29*, 2132.

[159]   G. Culcu, D. A. Iovan, J. P. Krogman, M. J. Wilding, M. W. Bezpalko, B. M. Foxman, C. M. Thomas, *J. Am. Chem. Soc.* **2017**, *139*, 9627.

[160]   D. A. Evers, A. H. Bluestein, B. M. Foxman, C. M. Thomas, *Dalton Trans.* **2012**, *41*, 8111.

[161]   P. P. Power, *Chem. Rev.* **1999**, *99*, 3463.

[162]   R. C. Fischer, P. P. Power, *Chem. Rev.* **2010**, *110*, 3877.

[163]   C. Ganesamoorthy, D. Bläser, C. Wölper, S. Schulz, *Chem. Commun.* **2014**, *50*, 12382.

[164]   B. Nekoueishahraki, H. W. Roesky, G. Schwab, D. Stern, D. Stalke, *Inorg. Chem.* **2009**, *48*, 9174.

[165]   N. Burford, P. Losier, A. D. Phillips, P. J. Ragogna, T. S. Cameron, *Inorg. Chem.* **2003**, *42*, 1087.

[166]   J. Vrána, R. Jambor, A. Růžička, M. Alonso, F. De Proft, L. Dostál, *Eur. J. Inorg. Chem.* **2014**, *2014*, 5193.

[167]   B. Prashanth, D. Bawari, S. Singh, *ChemistrySelect* **2017**, *2*, 2039.

[168]   L. Grocholl, I. Schranz, L. Stahl, R. J. Staples, *Inorg. Chem.* **1998**, *37*, 2496.

[169]   J. Prust, P. Müller, C. Rennekamp, H. W. Roesky, I. Usón, *J. Chem. Soc., Dalton Trans.* **1999**, 2265.

[170]   G. D. Frey, V. Lavallo, B. Donnadieu, W. W. Schoeller, G. Bertrand, *Science* **2007**, *316*, 439.

[171]   R. Azhakar, H. W. Roesky, J. J. Holstein, K. Pröpper, B. Dittrich, *Organometallics* **2013**, *32*, 358.

[172]   J. Grobe, S. Göbelbecker, B. Krebs, M. Läge, *Z. Anorg. Allg. Chem.* **1992**, *611*, 11.

[173]   C. Präsang, M. Stoelzel, S. Inoue, A. Meltzer, M. Driess, *Angew. Chem. Int. Ed.* **2010**, *49*, 10002.

[174]   L. Tuscher, C. Helling, C. Wölper, W. Frank, A. S. Nizovtsev, S. Schulz, *Chem. -Eur. J.* **2018**, *24*, 3241.

[175]   D. Matioszek, N. Saffon, J.-M. Sotiropoulos, K. Miqueu, A. Castel, J. Escudié, *Inorg. Chem.* **2012**, *51*, 11716.

[176]   D. Gallego, S. Inoue, B. Blom, M. Driess, *Organometallics* **2014**, *33*, 6885.

[177]   H. X. Yeong, S. H. Zhang, H. W. Xi, J. D. Guo, K. H. Lim, S. Nagase, C. W. So, *Chem. -Eur. J.* **2012**, *18*, 2685.

[178]   Y. Peng, B. D. Ellis, X. Wang, P. P. Power, *J. Am. Chem. Soc.* **2008**, *130*, 12268.

[179]   P. Wilfling, K. Schittelkopf, M. Flock, R. H. Herber, P. P. Power, R. C. Fischer, *Organometallics* **2014**, *34*, 2222.

[180]   J. K. West, L. Stahl, *Inorg. Chem.* **2017**, *56*, 12728.

[181]   M. Chen, J. R. Fulton, P. B. Hitchcock, N. C. Johnstone, M. F. Lappert, A. V. Protchenko, *Dalton Trans.* **2007**, 2770.

[182]   S. Hino, M. Olmstead, A. D. Phillips, R. J. Wright, P. P. Power, *Inorg. Chem.* **2004**, *43*, 7346.

[183]   A. Stasch, C. M. Forsyth, C. Jones, P. C. Junk, *New J. Chem.* **2008**, *32*, 829.

[184]   T. Chlupaty, Z. k. Padělková, F. DeProft, R. Willem, A. Ruzicka, *Organometallics* **2012**, *31*, 2203.

[185]   H. Vaňkátová, L. Broeckaert, F. De Proft, R. Olejník, J. Turek, Z. k. Padělková, A. Ruzicka, *Inorg. Chem.* **2011**, *50*, 9454.

[186]   W. A. Merrill, R. J. Wright, C. S. Stanciu, M. M. Olmstead, J. C. Fettinger, P. P. Power, *Inorg. Chem.* **2010**, *49*, 7097.

[187]   C. Stanciu, S. S. Hino, M. Stender, A. F. Richards, M. M. Olmstead, P. P. Power, *Inorg. Chem.* **2005**, *44*, 2774.

[188]   C. Cui, M. Brynda, M. M. Olmstead, P. P. Power, *J. Am. Chem. Soc.* **2004**, *126*, 6510.

[189]   M. Drieß, H. Pritzkow, *Chem. Ber.* **1994**, *127*, 477.

[190]   L. Zsolnai, G. Huttner, M. Driess, *Angew. Chem. Int. Ed.* **1993**, *32*, 1439.

[191]   Y. V. Fedotova, A. N. Kornev, V. V. Sushev, Y. A. Kursky, T. G. Mushtina, N. P. Makarenko, G. K. Fukin, G. A. Abakumov, L. N. Zakharov, A. L. Rheingold, *J. Organomet. Chem.* **2004**, *689*, 3060.

[192]   R. Aysin, L. Leites, S. Bukalov, A. Zabula, R. West, *Inorg. Chem.* **2016**, *55*, 4698.

[193]   S. Krupski, J. V. Dickschat, A. Hepp, T. Pape, F. E. Hahn, *Organometallics* **2012**, *31*, 2078.

[194]   J. Li, A. Stasch, C. Schenk, C. Jones, *Dalton Trans.* **2011**, *40*, 10448.

[195]   P. J. Davidson, M. F. Lappert, *J. Chem. Soc., Chem. Commun.* **1973**, 317a.

[196]   M. J. Gynane, D. H. Harris, M. F. Lappert, P. P. Power, P. Rivière, M. Rivière-Baudet, *J. Chem. Soc., Dalton Trans.* **1977**, 2004.

[197]   S. Krupski, R. Pöttgen, I. Schellenberg, F. E. Hahn, *Dalton Trans.* **2014**, *43*, 173.

[198]   J. P. Charmant, M. F. Haddow, F. E. Hahn, D. Heitmann, R. Fröhlich, S. M. Mansell, C. A. Russell, D. F. Wass, *Dalton Trans.* **2008**, 6055.

[199]   J. Volk, B. A. C. Bicho, C. Bruhn, U. Siemeling, *Z. Naturforsch. B Chem. Sci.* **2017**, *72*, 785.

[200]   F. E. Hahn, A. V. Zabula, T. Pape, A. Hepp, *Eur. J. Inorg. Chem.* **2007**, *2007*, 2405.

[201]   S. I. Al-Rafia, P. A. Lummis, M. J. Ferguson, R. McDonald, E. Rivard, *Inorg. Chem.* **2010**, *49*, 9709.

[202]   J. A. Cabeza, J. M. Fernández-Colinas, P. García-Álvarez, D. Polo, *Inorg. Chem.* **2012**, *51*, 3896.

[203]   J. Bareš, V. Šourek, Z. Padělková, P. Meunier, N. Pirio, I. Císařová, A. Růžička, J. Holeček, *Collect. Czech. Chem. Commun.* **2010**, *75*, 121.

[204]   J. P. Krogman, B. M. Foxman, C. M. Thomas, *J. Am. Chem. Soc.* **2011**, *133*, 14582.

[205]   S. Kuppuswamy, M. W. Bezpalko, T. M. Powers, M. J. Wilding, C. K. Brozek, B. M. Foxman, C. M. Thomas, *Chem. Sci.* **2014**, *5*, 1617.

[206]   J. Mautz, K. Heinze, H. Wadepohl, G. Huttner, *Eur. J. Inorg. Chem.* **2008**, *2008*, 1413.

[207]   Z. Dong, K. Bedbur, M. Schmidtmann, T. Müller, *J. Am. Chem. Soc.* **2018**, *140*, 3052.

[208]   D. Tofan, C. C. Cummins, *Chem. Sci.* **2012**, *3*, 2474.

[209]   P. Bazinet, G. P. Yap, D. S. Richeson, *J. Am. Chem. Soc.* **2001**, *123*, 11162.

[210]   S. Mitzinger, L. Broeckaert, W. Massa, F. Weigend, S. Dehnen, *Nat. Commun.* **2016**, *7*, 10480.

[211]   W. L. Armarego, *Purification of laboratory chemicals*, Butterworth-Heinemann, **2017**.

[212]   G. M. Sheldrick, *Acta Crystallogr. Sect. A: Found. Crystallogr.* **2008**, *64*, 112.

[213]   G. M. Sheldrick, *Acta Crystallogr. A* **2015**, *71*, 3.

[214]   O. V. Dolomanov, L. J. Bourhis, R. J. Gildea, J. A. Howard, H. Puschmann, *J. Appl. Crystallogr.* **2009**, *42*, 339.

[215]   G. Sheldrick, *Acta Crystallogr., Sect. C* **2008**, *41*, 3.

[216]   K. Diamond 4.3.2 - Crystal and Molecular Structure Visualization Crystal Impact - H. Putz & K. Brandenburg GbR, D-53227 Bonn.

8. Appendix

8.1 Directory of Abbreviations

2,2'-bpy	2,2'-bipyridine
Å	Angström (10^{-10} m)
Bz	benzyl
BTSA	bis(trimethylsilyl)amide
calcd.	calculated
COD	1,5-cyclooctadiene
Cp'	methylcyclopentadienyl
Cp	cyclopentadienyl
Cp*	pentamethylcyclopentadienyl
d	day(s)
Dipp	2,6-diisopropyl phenyl
DFT	density functional theory;
DCM	dichloromethane
E.A.	elemental analysis
EI	electron ionization
Et_3N	triethylamine
Et	ethyl
Eqn.	equation
Equiv.	equivalent
h	hour(s)
iPr	iso-propyl
IR	Infrared
KBTSA	potassium bis(trimethylsilyl)amide
LH	ligand
LK	potassium salt of ligand
Ln	lanthanide
M	metal
Me	methyl
MeLi	methyllithium
min	minute(s)
Mes	mesitylene or 1,3,5-trimethylbenzene
mg	milligram
mL	milliliter
mmol	millimole
MS	Mass Spectrometry
^{n}BuLi	n-butyllithium
NMR	nuclear magnetic resonance
p-Tol	4-methyl phenyl
Ph	phenyl

R	organic group
*t*Bu	*tert*-butyl
TM	transition metal
T	temperature
THF	tetrahydrofuran
Tol.	toluene, methylbenzene
TMS	trimethylsilyl
Ter	2,6-bis-(2,4,6-trimethylphenyl)
XRD	X-ray diffraction
Xyl	3,5-dimethylphenyl

8.2 NMR Abbreviations

MHz	mega hertz
dd	doublet
m	multiplet
qt	quintet
s	singlet
t	triplet
δ	chemical shift
ppm	parts per million

8.3 IR abbreviations

br	broad
m	medium
s	strong
vs	very strong
w	weak
vw	very weak

Acknowledgements

First, I would like to express my deepest gratitude and sincere thanks to my supervisor, Prof. Dr. Peter W. Roesky, for giving me many interesting projects and offering extremely helpful suggestions and encouragement during my PhD study. I really appreciate the time and effort he contributed to my study. I learnt so much not only from his profound knowledge, but also from his passion in chemistry and rigorous scientific approach. It is great luck to have joint his research group and working under his supervision. All of the help have been of inestimable worth to the completion of my thesis.

I would like to thank my parents for believing me and providing the endless support and love during my PhD. I would like to thank my relatives, especially my uncles and aunts, for their encouragement and support.

I would like to thank Ravi Yadav for his great help. As the closest friend and colleague, he gave me many good suggestions during my PhD.

I would like to thank Dr. Thomas Simler, Xiaofei Sun, Ravi Yadav, Benjamin Basler, Bhupendra Goswami, Mingting Hao and Christina Zovko for their good suggestions during my writing.

I would like to thank Sibylle Schneider for X-ray single crystal measurement and Dr. Michael Gamer for suggestions on crystallography problem.

I would like to thank Dr. Christoph Kaub and Sebastian Kaufmann for helping me with internet and software problems.

I appreciate Monika Kayas and Angie Pendl for their assistance with paper work.

I would like to thank Helga Berberich and Nicole Klaassen for the NMR and elemental analysis measurement.

I would like to thank all of my colleges for their help and suggestion during my work.

I thank my friends, Pingyin Guan, Dr. Gangliang Tang, Peng Xue, Kunxi Zhang, Ruipu Wang and Dr. Lixin Wang for their great company and encouragement during my stay in Karlsruhe.

Finally, I would like to thanks the China Scholarship Council (No. 201506250062) for providing the financial support for my PhD study.

Curriculum vitae

Name	Xiao Chen
Date of Birth	9. 10. 1990
Gender	Male
Place of Birth	Yong Cheng, Henan, China
Marital Status	Single
Nationality	China

Education:

Ph.D in Chemistry	(Sep. 2015 - present) Department of Chemistry, Karlsruhe Institute of Technology
Thesis Supervisor	Prof. Dr. Peter W. Roesky
Thesis Title	Synthesis and structural characterization of arsinoamide and its metal complexes
Master of Science	(Sep. 2012 - Jun. 2015) Tianjin University, Tianjin, China
Bachelor of Science	(Sep. 2008 - Jun. 2012) Jilin Institute of Chemical Technology, Jilin, China

Publications

1. Xiao Chen, Michael T. Gamer, and Peter W. Roesky. Synthesis and structural characterization of alkali metal arsinoamides. *Dalton Trans.*, **2018**, *47*, 12521-12525.

2. Meng He, Xiao Chen, Tilmann Bodenstein, Andreas Nyvang, Sebastian F. M. Schmidt, Yan Peng, Eufemio Moreno-Pineda, Mario Ruben, Karin Fink, Michael T. Gamer, Annie K. Powell, and Peter W. Roesky. Enantiopure Benzamidinate / Cyclooctatetraene Complexes of the Rare-Earth Elements: Synthesis, Structure, and Magnetism. *Organometallics*, **2018**, *37*, 3708-3717.

3. Xiao Chen, Thomas Simler, Ravi Yadav, Michael T. Gamer, and Peter W. Roesky. Reaction of an Arsinoamide with a Silylene and Germylenes: Substitution and As-N Bond Insertion. *In preparation.*